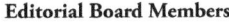

Published in the United States of America
by Michigan Publishing

DOI: http://doi.org/10.3998/gs

ISBN 978-1-60785-849-2 (paper, issue 3.1)

Front cover illustration by Ruepert Jiel Cao

Secretariat
The Academy of Film
Hong Kong Baptist University
CVA937, Lee Shau Kee Communication and Visual Arts Building, 5
Hereford Road, Kowloon Tong, Kowloon, Hong Kong.
Tel: (852) 3411 7493
Email: gstjournal@hkbu.edu.hk
Website: https://research.hkbu.edu.hk/project/global-storytelling-journal-of-
digital-and-moving-images

Supported by
The Academy of Film
The School of Creative Arts
Hong Kong Baptist University

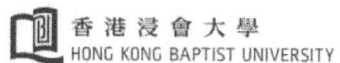

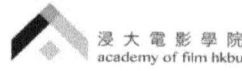

GLOBAL STORYTELLING:
JOURNAL OF DIGITAL AND MOVING IMAGES

Special Issue 3.1 – East Asian Serial Dramas
in the Era of Global Streaming Services
(Summer 2023)

Special Issue Editors: Tze-lan Sang, Lina Qu, and Ying Zhu

CONTENTS

Contents

East Asian Serial Dramas in the Era of Global Streaming Services

Special Issue Editors' Introduction

TZE-LAN SANG, LINA QU, AND YING ZHU

Keywords: East Asian serial dramas, SVOD platform, streaming, global cultural flows, IP

The story began in early 2022 when Tze-lan and Lina contacted Ying, the editor in chief of *Global Storytelling*, to propose a special issue. After brainstorming, we decided to have a three-way collaboration on the theme of East Asian serial dramas in the era of global streaming services and address the growing visibility and international circulation of East Asian cultural products.

Global streaming platforms have emerged as major outlets for the worldwide circulation of transnational programs, including East Asian serial dramas. Despite their tremendous success, the global consumption of East Asian dramas and their impact on storytelling, representation, and national image remain understudied. Many questions remain. For instance, how have global streaming giants influenced local serial drama production, viewer experience, and digital labor? In the era of global streaming, what is the role of the nation?[1] Do state policies—whether pertaining to the enforcement of

---

1. For a wide- ranging critique of platform imperialism, focusing on the role of the nation-state alongside transnational capital, see Dai Yong Jin, *Digital Platforms, Imperialism and Political Culture* (New York: Routledge, 2015).

https://doi.org/10.3998/gs.4292

international intellectual property laws, to political control, or to national branding—make a major difference? How do the television industries of different East Asian countries inspire one another while also competing for global ascendance? How does the proliferation of East Asian serial narratives on streaming platforms speak to international viewers' existing worldviews and complicate them?

For these reasons, we are especially keen on welcoming research that would address aspects of the interplay between local productions and global subscription video-on-demand (SVOD) platforms such as Netflix and Disney+ in the United States and iQiyi in China. In 2021, *Squid Game*, a South Korean dystopian television series created for Netflix, became an international sensation amid the COVID-19 pandemic and ranked as Netflix's most-watched show ever. In addition to rekindling the crave for K-dramas since the formation of the Korean Wave in the 2000s, the *Squid Game* phenomenon has put a spotlight on East Asian serial dramas on streaming platforms. The enormous success of *Squid Game* demonstrates a new future offered by global SVOD platforms to TV producers in East Asia. Meanwhile, East Asian producers must balance their pursuit of "global appeal" against local political, ideological, cultural, and economic concerns.

The emergence of SVOD has revolutionized television industries as new ways of watching TV are made possible by nonlinear media. The programming logics for Internet-distributed television have thus shifted from scheduling to curation, as a number of scholars have argued.[2] What and how catalogs of diverse content are curated and/or commissioned for production and how such curated programs are perceived globally and locally have emerged as central questions for industry practitioners as well as academic researchers. This special issue approaches these questions from the perspective of East Asian television studies. We ask how East Asian serial dramas are

---

2. Amanda Lotz, *Portals: A Treatise on Internet-Distributed Television* (Ann Arbor: University of Michigan Press, 2017); Ramon Robato, "Rethinking International TV Flows Research in the Age of Netflix," *Television & New Media* 19, no. 3 (2018): 241–56.

curated and/or commissioned for production by global SVOD platforms as well as investigate the interplay between the global streaming industry and the production, distribution, and reception of East Asian serial dramas.

Our call for papers was met with enthusiasm. After several rounds of screenings and balancing a number of factors, we present here five research articles, two drama reviews, and a short essay. Together, these papers highlight the vitality of East Asian cultural industries and the confluence of multiple forces: government policies, streamers' and media companies' commercial incentives and tactics, and audiences' experiences of not only consumption but also participation and interactive content creation. They draw a complex picture of how television has become global and how East Asian creative talents have generated sophisticated contents to appeal to not only local but international audiences.

David Humphrey delves into the intriguing question of why Japanese serial dramas are relatively scarce on global streaming platforms. In sharp contrast to the rising prominence of Korean dramas on the global scene, the export of Japanese dramas has stagnated in the past two decades. Employing a historical and comparative approach, Humphrey argues that the reason for the stagnation is structural rather than cultural. The dominance of broadcast television in Japan has hindered the industry's transition to streaming. The bundled-rights model, a hallmark of Japan's media industry since its adoption of the international intellectual property (IP) regimes in the 2000s, has limited the transnational and transmedia distribution of mainstream Japanese dramas while facilitating the export of anime and off-mainstream fare because of their lower IP hurdles. Humphrey also probes a conundrum facing Japan's IP-oriented media industry—that is, its increasing dependence on US-based streaming platforms and thus decreasing autonomy. His study underscores the friction between the local and the global in the international transmission of serial dramas: the localized rights and services of broadcasting are being challenged and reshaped by the decentralized streaming ecosystem.

Yucong Hao's article focuses on a new form of serial drama—the radio drama—against the backdrop of the Chinese state's tightening censorship of queer media. By tracing the history of queer media and *danmei* (Boy's Love) culture in China, Hao argues that the popularity of queer serial dramas on Chinese video-streaming platforms before the nationwide crackdown has paved the way to the emergence of queer radio dramas on podcasting platforms. The audio serial drama taps the cultivated queer viewership to gain popularity and market value while substituting queer visuals with voices to circumvent the state's stringent censorship. Building on current scholarships in queer and sound studies, Hao analyzes how the creative deployment of vocal timbre conveys queer motions and emotions in the radio drama *Grandmaster of Demonic Cultivation* (2016) and how such sound effects are collectively interpreted and instantaneously communicated by the online viewership via *danmaku* (bullet screens). The article provides a new angle on streaming media by presenting an intriguing case where voice streaming replaces video streaming to construct a queer mediascape at once elusive and tangible. It also sheds new light on serial storytelling by demonstrating how it can be enriched by active listeners' collective participation.

Winnie Yanjing Wu's article on the popular Apple TV+ series *Pachinko* (2022) reconceptualizes transnational TV through the prism of migration. As a hybrid, multicultural, and multilingual production that depicts the migrant life of a Korean family in Japan and the United States over eighty years, the huge success of *Pachinko* indicates the relevance of migration experience to global audiences in both literal and metaphorical senses. Through nuanced visual and linguistic analyses, Wu delves into the drama's depiction of migrants' quintessential struggles. What's more, she maintains that viewing the multilingual migration melodrama and navigating multiple subtitle tracks on the streaming platform is in itself a disorienting and disquieting experience, one that mirrors, to some extent, migrants' linguistic and cultural struggles in unfamiliar places. In other words, *Pachinko* allegorizes the impact of transnational TV on viewers, as the act of viewing transnational programs via streaming and the act of migration produce similar effects of time-space compression. Streaming platforms have transformed human experience and perception of time, space, language, homeplace, and identity just as migration has. Wu's refreshing approach contextualizes streaming media within the contested postmodern conditions. The saga of voluntary and involuntary migration illuminates the cultural politics of global streaming services.

Eunice Ying Ci Lim's study of the Taiwanese Netflix series *Mom, Don't Do That!* (2022) interprets nostalgic symbols in the show as self-reflexive commentaries on the rivalry between Taiwanese and Korean dramas. The dominance of Korean dramas in Asian and global entertainment scenes has concerned Taiwanese TV producers in recent years. Lim observes how the show's nostalgic representation of TV watching by ordinary Taiwanese families calls to mind the golden era of Taiwanese dramas in the 1980s and 1990s before the rise and encroachment of the Korean Wave. She also discerns a parodic strategy of representing food consumption and idealized masculinity in the show that critiques stereotypes in romantic Korean dramas. To read *Mom, Don't Do That!* as a metadrama by the Taiwanese TV industry, Lim highlights its undertaking of strategic moves to counterbalance the undeniable prominence of Korean dramas. The article contributes a new perspective to the study of streaming media with a focus on intra-Asia cultural flows and competition. It points out the uneven power dynamics in the global streaming economy while also suggesting streaming's potential to level the playing field.

While the above four research articles zoom in on East Asian serial dramas, Ying Zhu complicates the picture by examining Chinese fansub in the contexts of global cultural flows (from Hollywood to the Chinese Internet) and Chinese state censorship. As Zhu points out, the voluntary and collective labor of fansubbers has created a major channel for circulating authentic and unabridged foreign media content, particularly American films and serial dramas, in China. Although the legality of fansubbed audiovisual materials is questioned by international IP regimes, the political and cultural implications of fansubbing are crucial to the formation of a transgressive grassroots culture with therapeutic potentials in the Chinese cyberspace. As self-claimed cultural brokers, Chinese fansubbers are dedicated to the mission of introducing foreign cultures to Chinese audiences, whose access to foreign media content is hindered by IP laws and state censorship as well as linguistic and cultural barriers. At the same time, fansubbers undertake a self-healing journey to validate their own emotions and accomplishments through the practice of translating and subtitling. Zhu's work addresses an underresearched aspect of the international circulation of serial dramas and brings to the fore the active role played by media prosumers.

In addition to the five research articles, this special issue also includes two drama reviews and one short essay to provide a comprehensive approach to serial dramas and streaming services. The Korean Netflix series *Squid Game*, a phenomenal hit in 2021, is reviewed by Mei Mingxue Nan. Nan points out strategies used by Korean producers, the themes of neoliberal capitalism and constant surveillance, and the innovative camerawork (especially posthuman POV shots) in *Squid Game*. She also analyzes the issue of cultural appropriation and the flattening of local specificities in the process of packaging and selling East Asian stories for global consumption. Theorizing what she calls the "feeling of platform cosmopolitanism," she observes that many viewers praise the relatability of a series like *Squid Game* without any concern for the pitfalls of flattening. Paradoxically, this leads Nan to posit the possibility of a deeper engagement with the foreign Other, for platformization has enabled greater and faster knowledge-sharing for those who care to investigate and learn. In the nexus of platform-content-human, she discerns the potential of infra-individual intra-actions, to borrow Thomas Lamarre's theorization about platformativity.[3]

Shuwen Yang's review of *Light the Night*, another popular Netflix series in 2021, offers a detailed synopsis with useful background information about the historical references, locations, and actors in the Taiwanese drama. The show's global success supports Mei Nan's argument about platform cosmopolitanism as a prevailing cultural ideology of global streaming services.

---

3. Thomas Lamarre, "Platformativity: Media Studies, Area Studies," *Asiascape: Digital Asia* 4, no. 3 (2017): 285–305, https://doi.org/10.1163/22142312-12340081.

Finally, the short essay by Sheng-mei Ma parses the political irony and innuendo underlying an acclaimed crime TV series produced by the Chinese streaming giant iQiyi. For Ma, *The Bad Kids* is an example of what he terms "Red China Noir." The dark drama walks a tightrope between the government's upbeat rhetoric of the Chinese dream and the Chinese masses' fascination with crime and suspense in everyday entertainment. Using the Blakean dyad—the tyger and the lamb—Ma reads the horror and abyss lying just beneath the surface of childhood innocence in this drama. What does this series's acclaim and popularity tell us about the collective unconscious in China?

With a dedicated focus on East Asian audiovisual storytelling on streaming platforms, this collection of essays and reviews hopes to bring our attention to the growing influence and visibility of East Asian serial dramas during the era of platformization, and to (re)imagine transnational virtual storytelling from multiple geolinguistic, geocultural, geopolitical, and geo-economic persuasions and perspectives.

Acknowledgments

We would like to thank all the scholars who submitted their work for consideration. Thanks are also due the anonymous reviewers for each individual manuscript for their helpful comments.

Research Articles

The Therapeutic and the Transgressive

Chinese Fansub Straddling between Hollywood IP Laws and Chinese State Censorship[1]

YING ZHU

Abstract

Fansub has played a significant role in recent years in introducing otherwise unavailable foreign AV content to China via file transfer on pirate websites that bypass or play cat-and-mouse games with both regulators and copyright holders. Fansub further provides opportunities for participants to preserve the integrity of the source content that challenges mainstream conventions and values. This article provides an overview of fansubbing in China and discusses the complex issues involving international IP law, China's selective compliance with such laws, and the Chinese government's censorship of media and entertainment content. Specifically, the article traces the evolution of Chinese piracy of Hollywood film and television from counterfeit to fansub to tease out the larger issues of access, advocacy, and copyright infringement. Neither the legal nor the political hazard has deterred die-hard fansubbers from their "transgressions." In discussing fansub motivations, the article further examines both the transgressive and affirmative experiences of Chinese fansub through the lens of the therapeutic effects.

Keywords: Fansub, The therapeutic effects, IP laws, Censorship, Piracy

---

1. The author wishes to thank Dr. Xiqing Zheng, an assistant professor at the Institute of Literature, Chinese Academy of Social Sciences in Beijing, for sharing her thoughts and writings during the draft stage of this article.

Introduction

In February 2021, the Chinese government arrested more than a dozen people affiliated with the largest and most famous and popular Chinese piracy and subtitling site YYeTs.com (a site known domestically as Renren Yingshi), which was created in 2004 by a group of Chinese students in Canada. The arrests caused alarm among a community of people with shared passion for fansubbing (fansub), the process of fans translating via subtitling foreign audiovisual (AV) material without authorization into a local language—Chinese in this case—for free downloads. This article provides an overview of fansubbing in China and discusses the complex issues involving international IP law, China's selective compliance with these laws, and China's censorship of media and entertainment content. Given the popularity of US film and TV shows, the focus will be on Chinese fansubs' relationship with US entertainment content, which provides a case study to unpack some of these complex issues. Specifically, the article traces the evolution of Chinese piracy of Hollywood film and television from counterfeit to fansub to tease out the larger issues of access, advocacy, and copyright infringement. The article will further discuss fansub motivations and examine both the transgressive and affirmative experiences of Chinese fansub through the lens of the therapeutic effects, what I term the *therapeutic experience of fansubbing.*

Fansub has played a significant role in recent years in introducing otherwise unavailable foreign AV content to China via file transfer on pirate websites that bypass or play cat-and-mouse games with both regulators and copyright holders. Well-educated and mostly based in urban centers in and out of China, Chinese fansubbers come from a diverse range of white-collar professions, including engineers, accountants, university students and faculty, lawyers and physicians, as well as housewives.[2] As "self-appointed

---

2. Chi-hua Hsiao, "The Moralities of Intellectual Property: Subtitle Groups as Cultural Brokers in China," *Asia Pacific Journal of Anthropology* 15, no. 3 (May 2014): 218–41, https://doi.org/10.1080/14442213.2014.913673.

translation commissioners,"[3] fansubbers curate AV content for translation, with the earliest content coming from the United States and Hollywood films and TV dramas, which constitutes a significant portion of pirated material in the Chinese market. In addition to introducing new programs, Chinese fans of American films and TV dramas have also sought to monitor officially translated content and to safeguard the quality and integrity of the source content. There are many instances of staid official translation being outshined by the more vivid and engaging fan translation that utilizes collo-quial and vernacular Chinese "to closely render the meaning and register of the source-language dialogue," as enumerated by Dingkun Wang.[4] Fans also go out of their way to provide cultural and historical context in the form of glosses and notes.[5]

Fansub further provides opportunities for participants to preserve the integrity of the source content that challenges mainstream conventions and values. The award-winning film *Bohemian Rhapsody* (directed by Bryan Singer & Dexter Fletcher, 2018), for instance, was heavily reedited for its official China release, with six scenes of gay themes tossed out. But fansub offered a pirated source version, returning the full experience to Chinese viewers and winning grassroots endorsement. In such instances, fansub makes visible the otherwise invisible traces of censorship and its movements. Though winning grassroots support by exposing Chinese viewers to full and diverse viewing experiences, Chinese fansub groups exist in a legal grey zone due to the pirated nature of their practice, causing complaints from for-eign copyright holders for the loss of revenue and undermining the Chinese state in its international trade negotiations as the Chinese authorities seek

---

3. Luis Pérez-González, "Intervention in New Amateur Subtitling Cultures: A Multi-modal Account," *Linguistica Antverpiensia, New Series—Themes in Translation Studies* 6 (October 2021): 67–80, https://doi.org/10.52034/lanstts.v6i.180.

4. Dingkun Wang, "Fansubbing in China—with Reference to the Fansubbing Group YYeTs," *Journal of Specialised Translation Issue* 28 (2017): 165–90, https://www.jostrans.org/issue28/art_wang.pdf.

5. Tessa Dwyer, "Fansub Dreaming on ViKi," *Translator* 18, no. 2 (2012): 217–43, https://www.tandfonline.com/doi/abs/10.1080/13556509.2012.10799509

to selectively comply with IP laws. Chinese fansub exists in a political grey zone for its transgressive practice that bypasses Chinese censors who control the inbound foreign contents.

Neither the legal nor the political hazard has deterred die-hard fansubbers from their "transgressions." How does one account for the persistence of fansub in China? In the face of inferior or incomplete translations as the result of either the stilted language of official translation or translations with intentional omissions due to the government's surveillance of sensitive content, fansubbing helps the fan community to bypass the shoddy "official" channels for a more authentic encounter with the source content. But the access fansub offers is not entirely unfettered or undiscriminatory, as the process entails content curation based on a fan's assessment of the worthiness of source contents for translation. How do Chinese fansubbers determine the materials they wish to translate? What motivates them to put in the free labor for such an endeavor? What makes the endeavor pleasurable or therapeutic? One way to understand the motivations of fansubbers is to examine the actual experiences of fansubbing. As described by fansubbers I have encountered, the process of translating and sharing popular entertainment contents can trigger instant and, at times, supercharged corporeal and emotional reactions among the fandom, which are central to our sensory engagement with audiovisual content as well as the world, and which form part of the affective therapeutic experience. The article ventures two possibilities of fansubbing motivation: a deeper understanding of the text through the labor of translation and the recognition and affirmation from the fan community.

Fansub, Piracy, and the Demand for American AV Contents

Fansubbing has become a crucial form of alternative distribution for foreign language AV content all around the world. In China, fansubbing includes media products of diverse national and linguistic origins, from Thai drama to Ukrainian documentaries, but the most well-known fansubs in China has been those of Hollywood films, English-speaking TV dramas, and Japanese anime and dramas. AV content originating from the United States occupies a large share of the Chinese fansub repertoire. Volunteer subtitle groups emerged as early as the late 1990s, largely to satisfy demand among Chinese youth for an authentic and unadulterated viewing experience of popular US entertainment programs.

Fansubbing of American television series began to snowball in 2003 when the sitcom *Friends* was introduced to China via online streaming and pirated DVDs.[6] In response to the popularity of the series, an online forum, F6, was founded to provide fansubbing, and the term *měi-jù* (美剧, American television series) became a buzzword. Fansubs of American AV content were widespread between 2003 and 2005, causing considerable consternation among the US copyright holders who saw revenues from official distribution dissipate, resurrecting Hollywood's concern for piracy, which was "one of the thorniest issues in the olden days of Sino-Hollywood negotiation."[7] Indeed, prior to the era of streaming, the limited access to Western movies and TV shows had led to burgeoning demand for pirated AV contents, chiefly Hollywood films and TV dramas, making China one of the world's most prolific audiovisual counterfeiters of VHS in the 1980s and VCD and DVD in the 1990s.

As detailed in chapter five of Zhu's book, *Hollywood in China: Behind the Scenes of the World's Largest Movie Market*, the United States began pressuring China to adopt stringent intellectual property laws to protect IP rights as soon as China opened its door to foreign imports, opening a floodgate of pirated Hollywood films while limiting the number of titles in official circulation.[8] In 1988, the US Congress passed Special 301 of the 1988 Trade Act, giving the United States an effective tool to deal with nations that

6. Wang, "Fansubbing in China," 168.
7. Ying Zhu, *Hollywood in China: Behind the Scenes of the World's Largest Movie Market* (New York: New Press, 2022), 160
8. Zhu, *Hollywood in China.*

imposed barriers against US film and entertainment while also forbidding piracy of US audiovisual products. Countries identified by the US trade representative (USTR) under Special 301 could face a variety of retaliatory actions. Those with the greatest potential for adverse impact on US products were designated as "Priority Foreign Countries," which were subject to trade sanctions. In addition to the list of Priority Foreign Countries, "Priority Watch List" and "Watch List" were lesser categories that would not incur immediate trade sanctions. The USTR placed China on the Priority Watch List in 1989 and 1990 consecutively as it coaxed China to pursue intellectual property rights (IPR) legal reform.

Under the pressure, China, in 1990, promulgated the first copyright law under the PRC, which came eighty years after China's last dynasty issued the nation's first copyright law, the "Qing Copyright Code" in 1910.[9] The 1910 Copyright Code of the Great Qing Dynasty was short-lived, as the 1911 Revolution led by Dr Sun Yat-sen soon overthrew the Qing Dynasty. Though the idea of intellectual property was fundamentally at odds with Chinese tradition, the Qing Copyright Code nevertheless influenced copyright laws in China during the Republic era. But the People's Republic during Mao's era had little patience for copyright, and indeed intellectual property protection in general. It was not until 1990 that the Chinese Communist Party (CCP) established China's first copyright law. But the PRC copyright law signed in 1990 did not adhere to the Berne Convention, an international agreement governing copyright adopted in Berne, Switzerland, in 1886. Under the pressure from the United States, China pledged, in January 1991, to join the Berne Convention and adhere to the Geneva Phonograms Convention within the following two years, agreeing to make US products, including Hollywood films, "fully eligible for protection."

Piracy of Hollywood films ran so rampant in China by the early 1990s that it posed a major challenge to Hollywood's revenue. The Motion Picture

---

9. Yiping Yang, "The 1990 Copyright Law of the People's Republic of China," *UCLA Pacific Basin Law Journal* 11, no. 2 (1993), https://doi.org/10.5070/p8112022041.

Association of America (MPAA) eventually opened its Beijing office in 1994, with the twin priorities of ensuring China's effective enforcement of IPR while pressing for greater market access for major Hollywood studios, which Hollywood saw as crucial to combat piracy. In the MPAA's view, Chinese piracy was driven by censorship, quota barriers, and delayed distribution of Hollywood films. The proposed solution was for China to allow greater official distribution of Hollywood productions to mitigate the problem of piracy.

By the mid-1990s, roughly two hundred domestic films and sixty imports were released annually in China, but US films only took up ten slots among the sixty foreign titles despite the popularity of US entertainment products among Chinese audiences, which continued to incentivize piracy.[10] The MPAA took the initiative to directly engage legal and investigative companies in China to track down and carry out raids on pirates. In August 2003, Twentieth Century Fox, Disney, and Universal Studios won a civil lawsuit in Shanghai against two local companies selling pirated DVDs of Hollywood films. Yet film piracy continued to rage in China as Hollywood products became increasingly popular, making the low-capital piracy business an ever more lucrative enterprise—anyone could afford to counterfeit, and no one in China considered the sale of counterfeited goods a serious crime. As pressure mounted from the US side, Wu Yi, then Chinese vice premier and former head of the Ministry of Foreign Trade and Economic Cooperation, announced in 2004 that China would seek to lower the criminal threshold for piracy while also increasing the number of infringing acts that were subjected to criminal penalties. The same year, the Chinese regulator tightened its censorship control by issuing guidelines banning foreign programs deemed offensive to the Chinese sensibilities and disruptive to China's social stability.[11]

---

10. Ying Zhu, "*Hollywood in China.*"
11. For a detailed list, see Dingkun Wang and Xiaochun Zhang, "Fansubbing in China," *Target: International Journal of Translation Studies* 29, no. 2 (2017): 301–18, https://doi.org/10.1075/target.29.2.06wan.

As Hollywood fought hard against tangible counterfeits of optical disks, nontangible piracy quietly emerged online in China in the early 2000s by riding the wave of peer-to-peer (P2P) file-sharing. This new breed would transform the infrastructure of media piracy "from the industry-organized, commercial manufacture of optical disks to user-generated, peer-to-peer content sharing on computer networks."[12] The rapid development of the Internet and the reduction in the price of computers in the early 2000s allowed easier and greater access to foreign AV products. Before the Internet age, fansubbers used a complicated technical procedure to type and record on VHS tapes. A device called Genlock could synchronize two different video signals, enabling the fansubbers to add subtitle translations to the illegally distributed videotapes. Fans would buy these tapes from underground clubs.[13] P2P websites such as eMules.com made it easy for fans to download foreign films and TV dramas.

But fansubbing is illegal under the international IP legal framework, and the US annual "Special 301 Report" on copyright violation makes no exception for fansubbing. Though many foreign media products have circulated in China as pirated versions, some with subtitles translated by TV stations or DVD makers in Taiwan or Hong Kong, only the popular contents from major US production companies have captured widespread media interest. It was the fansubs of popular American TV shows that first attracted media attention during the era of online piracy. In August 2006, the *New York Times* broke a story about Chinese subtitle groups translating and making available to Chinese audiences unauthorized US TV dramas such as *Lost*, *C.S.I.*, and *Close to Home*.[14] Chinese fansubs' sudden US media

---

12. Jinying Li, "Pirate Cosmopolitanism and the Undercurrents of Flow Fansubbing Television on China's P2P Networks," in *Transnational Convergence of East Asian Pop Culture*, ed. Dal Yong Jin and Seok-Kyeong Hong (Abingdon and New York: Routledge, 2022), 127–46.

13. Marc Shaw, "How VHS Tapes and Bootleg Translations Started an Anime Fan War in the 90s," Vice, April 22, 2017, https://www.vice.com/en/article/pg5gqk/anime-fushigi-yugi-fan-subtitles-nineties-ottawa-cosplay-vhs.

14. Howard W. French, "Chinese Tech Buffs Slake Thirst for U.S. TV Shows," *New York Times*, August 9, 2006, http://www.nytimes.com/2006/08/09/world/asia/09china.html?_r=2&.

attention was partly the result of fansub's encroachment of popular main-stream entertainment programs instead of nonmainstream AV contents that belong to the so-called geek canon with mostly "nerdy" followers, which has been overlooked by the mainstream media.

Following the *New York Times*'s story, the "underground" circulation of the US serial drama *Prison Break* was reported by the Chinese mainstream media in late 2006, though no legal actions were taken by the Chinese government. While Hollywood complained of copyright violations, the Chinese government had been rather lenient initially at cracking down on fan activities "as long as everything is kept at the material consumption level and within the party line," to quote Weiyu Zhang and Chengting Mao.[15] Chinese copyright law permits personally produced media, which are defined as nonprofit-oriented and are shared only among friends. Fansubs in the name of sharing experience and knowledge of foreign languages was considered legitimate and at times even encouraged, leaving room for Chinese subtitle groups to thrive.

In 2007, two weeks before its official release in the United States of *Spider-Man 3*, subtitled DVDs supposedly containing the latest Hollywood blockbuster movie were spotted for sale on Beijing streets, reigniting Hollywood's call for piracy crackdown.[16] In a trip to Shanghai in June 2009 to attend the Shanghai International Film Festival, Dan Glickman, the head of the MPAA, complained that the growth of film piracy was costing studios billions each year in potential revenue. Glickman lobbied China's domestic content providers for a common effort to swiftly remove pirated online content. Legalization of online media via authorized domestic sites gradually arrived in China throughout the 2010s, the period when foreign producers started to sell rights for online streaming to Chinese media companies,

---

15. Weiyu Zhang and Chengting Mao, "Fan Activism Sustained and Challenged: Participatory Culture in Chinese Online Translation Communities," *Chinese Journal of Communication* 6, no. 1 (March 2013): 45–61, https://doi.org/10.1080/17544750.2013.753499.
16. See chapter six of Ying Zhu, "*Hollywood in China.*"

making Chinese domestic distributors Hollywood stakeholders for shared revenues. Though most fansub content did not have a legal venue to enter the Chinese market, Chinese fansubbed materials were easily accessible to audiences, sometimes even overshadowing legally imported media products. By the mid-2000s, fansubbed foreign films, TV shows, and anime could be found directly on Chinese video streaming websites, including Youku, Tudou, Ku6, ACFun, Bilibili.

Web 2.0 has made it possible for fans to connect directly with content producers, with some even participating in the "official" production process. Fansubbing groups subsequently cooperated with licensed domestic video-on-demand (VOD) websites to translate copyrighted foreign content. YYeTs, for example, was contracted by the video-streaming platform Sohu in 2010 to produce subtitles for the US show *Lost* (2004–2010) (Figure 1). This practice recalls an earlier era when Hollywood used the same tactics to co-opt counterfeiters by recruiting former pirators as their well-connected licensees for local distribution in China. Xianke, a pirated-copy distributor, was sued by the MPAA in the Chinese courts in 1994 and ordered to compensate the MPAA for damages and lawyers' fees as well as court expenses. Two years later, Warner Brothers turned around and appointed Xianke as an official distributor.[17] By converting sophisticated and efficient piracy networks to legitimate distribution channels, Hollywood majors managed to co-opt its illicit competitors, minimize financial costs, and mitigate losses.

Meanwhile, the Chinese government launched numerous anti-piracy campaigns that targeted P2P networks throughout the 2010s.[18] The campaign has continued into 2020s. Curiously, as noted by Li Jingying, the lead agency in cracking down P2P networks in the 2010s was neither the National Copyright Administration (the office responsible for copyright protection) nor the Ministry of Industry and Information Technology (the major administrative body regulating the Internet), but rather the

---

17. Ying Zhu.
18. Li, "Pirate cosmopolitanism," 128.

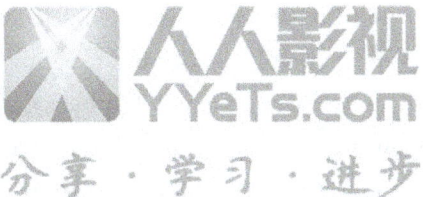

Figure 1: YYeTs' original logo and slogan: "Share, Study, Progress."
Source: YYeTs' Sina Weibo

State Administration of Radio, Film and Television (SARFT), the executive branch responsible for censoring media content under the direct order of the Ministry of Propaganda. As Li Jingying succinctly put it, the viral, distributive, and infiltrating online structure of the peer-to-peer network alarmed Chinese censors for its ability to instantly circulate content deemed inappropriate by the Chinese government. What began as a financial and IP issue has now been amplified into a political challenge of unfettered transnational cultural flow. It is worth noting that SARFT was merged with the General Administration of Press and Publication to form the State Administration of Press, Publication, Radio, Film and Television (SAPPRFT) in 2013. SAPPRFT was replaced in 2018 by the National Radio and Television Administration (NRTA) under the direct control of the CCP's Central Propaganda Department, further strengthening content regulations.

Unlike the earlier profit-driven counterfeiters, fansubbers view themselves as modern-day Robinhoods who operate in the spirit of volunteerism and free-sharing with the mission of promoting and making available otherwise censored foreign AV content. Major subtitle groups such as YYeTs run their own website and video-streaming mobile app, accompanied with an e-dictionary to facilitate viewing and enhance Chinese fans' foreign media literacy. Li Jingying's framework of a fansub group identity as "pirate cosmopolitanism" captures well fansubbers' self-image as sophisticated cultural brokers who subvert both the for-profit capitalist system and the mind-control authoritarian regime.[19] In a similar vein, Chi-hua Hsiao, writing in 2014, deemed Chinese fansub "a moral enterprise" that delivers public goods through "disciplined practice, volunteer work and devotion to the media programs."[20] In availing their language skills and technical know-how, fansubs curate for Chinese viewers what they see as worthy foreign AV content. On World Intellectual Property Day in 2013, Chinese authorities shut down some subtitle websites, leading to online protests by fansubbers who compared themselves to Prometheus, equating pirates with those who steal fires to (en)lighten (盗版者就是盗火者).[21] This view was echoed by many Chinese netizens who perceive fansub groups as rebels with a just cause.

US copyright holders initially ignored fansub, as the US content producers and Chinese distributors saw Chinese fansub during its earlier stage as a testing ground for potential importation through China's official channels.[22] Fansub indeed provided valuable market tests for international content producers, and authorized domestic VOD websites that distributed foreign contents. But Hollywood's initial forbearance eventually gave way to vigilance as Chinese fansubs proliferated. In October 2014, the MPAA singled out YYeTs.com as among the world's most pernicious sources of online DVD piracy for offering Chinese subtitles for unlicensed Western content, many of which were US-made movies and TV shows. YYeTs.com temporarily went offline in November 2014

---

19. Li, 127.
20. Hsiao, "The Moralities," 220.
21. Hsiao, 226; Chi-hua Hsiao, "The Cultural Translation of US Television Programs and Movies: Subtitle Groups as Cultural Brokers in China" (PhD diss., University of California, Los Angeles, 2014).
22. Lara Farrar, "Found in Translation: China's Volunteer Online Army," CNN, accessed March 30, 2023, http://edition.cnn.com/2009/BUSINESS/06/15/china.underground.translate/.

to "clean up" its content but only to reappear in 2015, at one point moving its servers to South Korea.[23]

China issued a regulation in early 2015 to tighten control over online streaming of foreign movies and TV dramas on authorized VOD websites. Starting from March 2015, all foreign movies and TV shows to be streamed on video sites must register with the media authorities and obtain a license. All content must be screened before broadcast to prevent unexpected stories and dialogues that touch sensitive issues. Quotas were set to cap foreign movies and TV shows to less than 30 percent of the total number of domestic contents the streaming sites broadcast in the previous year.[24] The broadcast time of imports must not exceed 25 percent of each site's daily schedule. With less and limited content, official VOD websites soon lost their advantages to fansubbing websites, prompting calls for legal enforcement from the copyright holders, especially the local representatives of foreign copyright holders who nudged the Chinese regulators to act. The 2020s has seen the shutdown by the Chinese government of over 2,800 websites and apps offering pirated content and the deletion of 3.2 million links.[25] The crackdown would eventually bring down YYeTs, which reportedly offered 32,824 unauthorized film and television shows to an estimated 6.83 million members by 2021. The prosecution of YYeTs under the supervision of China's National Copyright Administration, National Anti-Pornography Office, Ministry of Public Security, and the Supreme Procuratorate revealed that the company collected membership/subscription fees ("donations") while

23. Alexa Olesen, "A Mournful Farewell to Chinese Copyright Pirates," Foreign Policy, November 25, 2014, https://foreignpolicy.com/2014/11/25/a-mournful-farewell-to-chinese-copyright-pirates/.

24. Laney Zhang, "China: Control over Foreign Movies and TV Shows on Online Video Sites Tightened," Library of Congress, September 2014, https://www.loc.gov/item/global-legal-monitor/2014-09-22/china-control-over-foreign-movies-and-tv-shows-on-online-video-sites-tightened/.

25. According to the most recent data available from "NCAC-Top News," National Copyright Administration of China, accessed March 31, 2023, https://en.ncac.gov.cn/copyright/channels/10361.shtml.

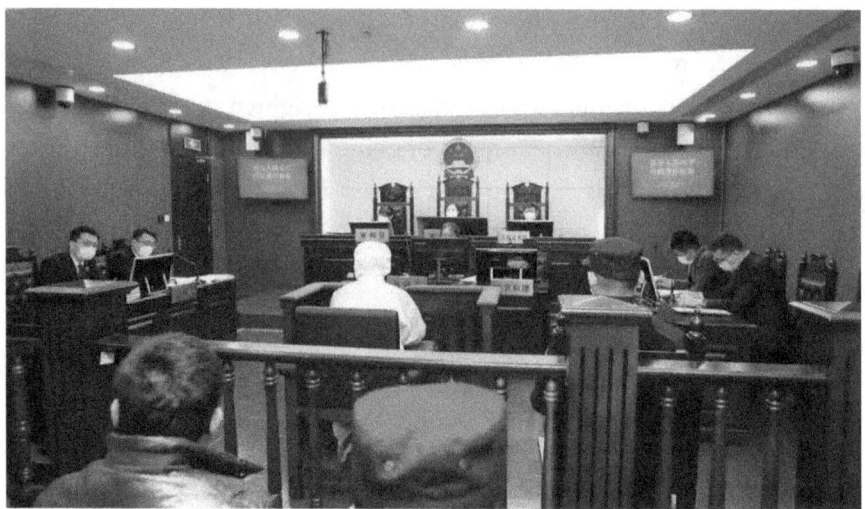

Figure 2: Photograph of Liang Yongping's first trial on November 21, 2021, when he was charged with three-and-a-half-year prison sentence and a fine.
Source: Sina Weibo

generating revenue from ads as well as hard-disk sales of unauthorized movies and TV shows since 2018, with the total amount of illegal revenue exceeding RMB 12 million ($1.87 million USD). These profit-making activities were deemed to have violated the Chinese law that prohibited profit-making copyright infringements. It soon followed that the founder Liang Yongping pleaded guilty and was handed a three-and-a-half-year prison sentence and a fine of RMB 1.5 million ($235,000 USD) (Figure 2).

The World of Chinese Fansub and the Therapeutic

One difficulty with many media products, especially of cult fandom and geek canon such as *Lord of the Rings*, is the overt complexity of story settings, which presents challenges for viewers to master all the historical and culture details in a convoluted story world. Lack of cultural background and widespread misinterpretation further complicate the translation of Western fantasy, which is immensely popular in the Chinese market. While fans consume and digest media and literary products repeatedly with full nerdy enthusiasm, local commercial advertisers and agencies do not share the same incentive to acquire and provide the comprehensive background knowledge in the fictional Middle Earth. Inferior or inaccurate translations by the official domestic distributors have been frequently ridiculed by the fan community with far deeper knowledge, understanding, and familiarity with the source product.

One tendency in recent years is for fansub groups to cluster around a singular object of interest—*Star Trek*, for instance. The major components in the Chinese fan reception of *Star Trek* include (1) the text (or the metatext as envisioned by the fans); (2) the "true" producers (usually referring to Gene Roddenberry and those who truly represent the ideal for the series and the films); (3) the "real" producers, including the director J. J. Abrams, the film studio, the screenwriters, etc.; (4) the domestic (or official, or commercial) representative of the producers, which is the film company that imports the film, the theaters that screen it, and the people who do the translation and the promotional activities; (5) Trekkies in US and other English-speaking countries, who are large in number and assume the position of authority from their celebration of the original creator of *Star Trek*, Gene Roddenberry, and the "spirit" they get from the original series; (6) Trekkies in China, who are in spirit the true companion to US Trekkies; and (7) the average moviegoers who have little idea of either the *Star Trek* universe or the original series. There are two lines that are drawn here: one is between the "official" and the text/fans, the other between English and Chinese. The first is easily discerned: for fans, original texts belong to them; foreign fans are their allies; the "true" producer and his spirit are always on their side (he is dead, anyway); the "real" producers, in this case, are not; the domestic representatives are worse. The second one is interesting, because, on the commercial side, there are authorities in both the Chinese and English environment, but on the textual side, two distinct figures contend to assume the role that mediates between Chinese and English: Chinese fansubbers and domestic official representatives. They represent two types of translation and intercultural negotiation—one of the fans, the other representing commercial interests.

Chinese Trekkies attain the most authority in the fandom as they interact with the source text directly, much like the US Trekkies do, and have thus become trustworthy representatives of the original text if a new Trekkie in China wants to learn more about the world of *Star Trek*. As noted in Zhang Xiaochun,[26] most Chinese fansubbers consider themselves cultural brokers between China and the world. Both sanitized content due to censorship and incompetence of translation on official channels provoke indignation among fans with abundant knowledge, which only compels fans to continue their practices, never mind the legal or potentially political persecutions.

While analyzing fansubs of Japanese anime, Ian Condry, among others, observes that there is an ethical code in the English-speaking fandom for Japanese anime, which stipulates that fan translators are to remove their translations of the original products once an entity legally imports and translates them.[27] Even though fans believe in the superiority of their translations, they consent to protect the IPR of the original producers and legal importers. Chinese fansubbers, however, largely do not honor the same ethical codes. Chinese fansubs are not incentivized to remove their own translations even after the product is officially imported through legal venues. In fact, many Chinese fansubbers would intentionally, indeed defiantly, retranslate their favorite content as a showcase for better and more sophisticated translations. One explanation points to China's lack of an IPR tradition and awareness, as Chi-hua Hsiao reminds us that China during

---

26. Xiaochun Zhang, "Fansubbing in China," Multilingual, July 1, 2013, https://multilingual.com/articles/fansubbing-in-china/.

27. Ian Condry, "Dark Energy: What Fansubs Reveal about the Copyright Wars," *Mechademia* 5, no. 1 (2010): 193–208, https://muse.jhu.edu/article/400557.

the Imperial era had a checkered history prioritizing information control at the cost of protection of property rights for individual authors and inventors.[28] This legacy has continued during the PRC era. Pirated and smuggled media products were rampant in the Chinese market before Chinese fansubbers emerged online, nurturing a generation of cinephiles turned fansubbers who are accustomed to free and easy access to foreign content with little consideration for regulatory and ethical issues related to international copyrights. Even as some of the most popular websites were taken down, fansubbing is still operating online, though some do try to abide by IPR by providing subtitles separately from the video, which means that the process of translating and sharing subtitles no longer involves disseminating the unlicensed original videos.

Scholars have argued that fansub subverts state censorship. They see the act of fansubbing as political activism and civic engagement that inspires Chinese viewers to confront state power and official oppression.[29] While contempt for state censorship might be one motivating factor, I'd venture a less explored aspect of fansub in China. I propose that Chinese fansub could be driven equally by the need for personal fulfillment and gratification, particularly and precisely reacting to the squelching of dissent by China's political system. As Laurie Cubbison points out, fans are motivated by the urgency to experience authentic content.[30] Fansubbing can be therapeutic as the process of producing subtitles grants fansubbers an outlet to channel their creative energy with transgressive usages of Chinese terms and phrases in their translation. Some fans actually rewrite the originals in a tongue-in-cheek fashion, appropriating colorful Chinese idioms to inject local issues into the original dialogues. By improvising via localization in translation,

---

28. Hisao, "The Moralities," 227.
29. Melissa M. Brough and Sangita Shresthova, "Fandom Meets Activism: Rethinking Civic and Political Participation," *Transformative Works and Cultures* 10 (March 2011), https://doi.org/10.3983/twc.2012.0303. Also in Wang and Zhang, "Fansubbing in China."
30. Laurie Cubbison, "Anime Fans, DVDs, and the Authentic Text," *Velvet Light Trap* 56, no. 1 (2005): 45–57, https://doi.org/10.1353/vlt.2006.0004.

Chinese fansub substitutes literary meanings in the source text with colloquial expressions to highlight local issues and to vent their frustration with the translation establishment, both government and corporations.

The widespread phenomenon in fansub of Tu Cao (吐槽), which means to roast or ridicule, allows translation to deviate from the source text, some with added notes and glosses that go beyond explaining cultural references to register the translator's feelings and sentiments on a cluster of social issues, not least of which is censorship, which fansub bypasses with levity and creativity. Among the taboos in Chinese culture, sexually charged words are particularly tricky. While euphemism reigns in translation, fansubbers can get creative. As mentioned in Zhang Xiaochun, "Jack, slow f**k" in *Titanic* was translated in one fansubbed version as "Czechoslovakia," which in Chinese is pronounced as *jié kè sī luò fá kè*, which is phonetically similar to "Jack, slow f**k" (Figure 3).[31] Venting their frustrations by poking fun at taboos and highlighting current events with creative translations, fansub has turned the translation traditionally associated with an elite occupation with stodgy language into a grassroots therapeutic exercise embellished with colloquial expressions and social commentaries. Through individualized self-expression, fansub on streaming has become another social media platform that facilitates emotional relief, transforming the intense labor of translation into a process of self-actualization while also empowering others. Zhang Xiaochun notes further that, while explaining his motivation, a translator known for his work on *Prison Break* said simply, "I love, so I share."

Fansubbers derive pleasure from the action of sharing their free labor for the common good. As Roger Foster puts it, the promise of the therapeutic culture is self-fulfillment rooted in the "search for the 'true' self," or "authenticity."[32] The therapeutic effect is augmented by participating in a

---

31. Zhang, "Fansubbing in China," 2013.
32. Roger Foster, "Therapeutic Culture, Authenticity and Neo-Liberalism," *History of the Human Sciences* 29, no. 1 (February 2016): 99–116, https://doi.org/10.1177/0952695115617384.

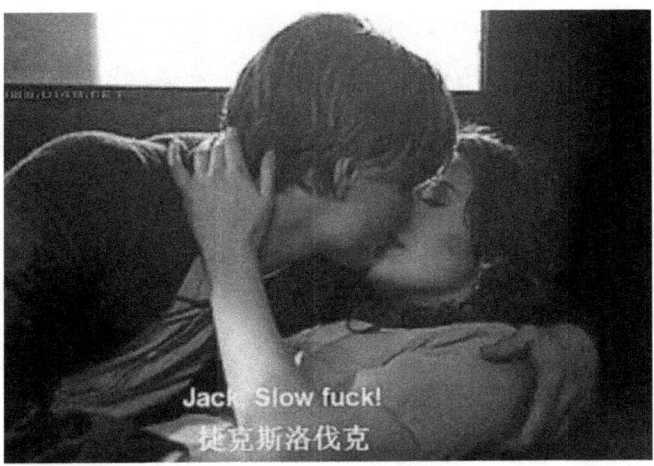

Figure 3: Fansubber's translation of "Jack, Slow Fuck" in *Titanic*.
Source: Zhihu

networked community of fansub with shared beliefs, behaviors, and mis-
sions that provide safety and security for fansubbers to share emotions and
translation antics. As Dhiraj Murthy argues, the positive feedback of one's
virtual peers is one of the major attributes of the therapeutic effects, which
brings recognition and confers social capital, two crucial components in the
self-actualization of individuals.[33] In an interview with the Guangzhou-based
Southern Metropolis Daily, Chinese fansubbers talked about how seeing one's
own name appearing in the opening credits of a completed translation work
and the wide circulation of such work could bring a sense of pride and
accomplishment.[34] There also exists hierarchies among fansub group members based on each individual's language proficiency and fan endorsement,
which encourages fansub ranking.[35] Though bringing no monetary rewards,

33. Dhiraj Murthy, "Twitter: Microphone for the Masses?" *Media, Culture & Society* 33, no.
    5 (June 2011): 779–89, https://doi.org/10.1177/0163443711404744.
34. Nanfang Dushi Bao, "Zimuzu Fanyi de Dashenmen (the Gods of Fansubbing Groups),"
    Sohu, August 6, 2014, http://news.sohu.com/20140806/n403155787.shtml.
35. Wang, "Fansubbing in China," 177.

fansubbers can nonetheless accrue social capital by generating followers and expanding their reach, compelling them to engage further in the fansub activity, making fansub an addictive experience akin to the addictive viewing experience of binge-watching via streaming of serial dramas, which the Chinese fans actively translate and share.

As more subtitle groups emerge, the competition for fast and efficient delivery and quality experience has led to parallel bilingual subtitles superimposed at the bottom of the videos, with some providing extensive explanatory notes on idioms, cultural references, and historical practices, all competing for eyeballs. Chinese fan-streaming apps further provide "live-streaming" translation service wherein translation takes place in real time, with only a few minutes' delay, which allows the fan community to enjoy the newly released show with only a short time lapse, leaving room for fans to participate in online discussions. The intense labor and time commitment involved in live streaming attests further to the linkage between the therapeutic and the addictive in the practice and experiences of fansubbing in China, which is not dissimilar to the effects of "morphine drip" in Denise Broe's description of the binge-consumption of serial TV dramas on demand (Figure 4).[36]

Concluding Remarks

Residing in a legal forbidden zone as transgressors of both international copyright laws and the domestic censorship regime, Chinese fansubbers ironically owe the relevance of their very existence to continuing state censorship, which forbids unfiltered foreign content from flowing into the Chinese market through legal channels. While IPR matters to foreign copyright holders and their official Chinese content distributors, for the Chinese government, the loss of cultural control poses an even greater threat

---

36. Dennis Broe, *Birth of the Binge: Serial TV and the End of Leisure* (Detroit, MI: Wayne State University Press, 2019).

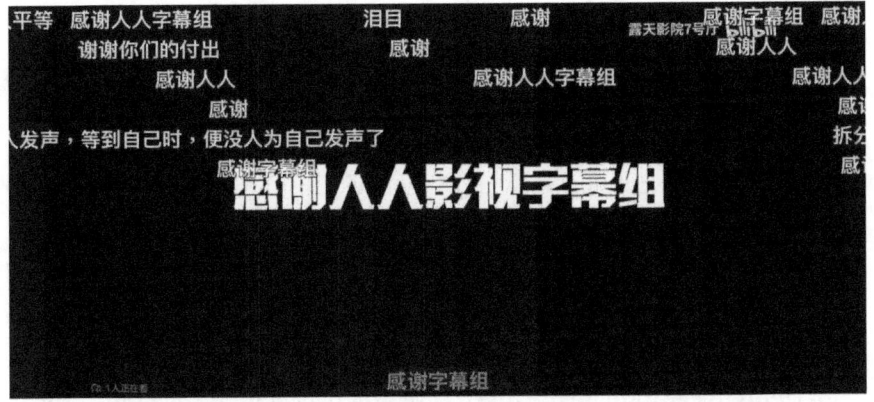

Figure 4: After YYeTs was completely cracked down in February 2021, a fan/Bilibili user made the video, "I am not the God of Dramas" (*wo bu shi jv shen*) to reminisce the disappearance of the platform. Other fans also acknowledged YYeTs in bulletin comments.
Source: Lutian Yingyuan Qihaoting on Bilibili

than copyright infringement. But the notoriously high piracy rate in China brings challenges to China's global standing in the IP industry amid the Chinese government's push to transform "made in China" to "created in China." Piracy is easily understood as a sign of insufficient creativity and originality, two keywords in the hierarchies of a global creative economy. Symbolically linked with knowledge and creativity, the strength of IP has become an indicator of a country's pecking order in the creative economy, which China is eager to harness.[37] It remains to be seen if the Chinese government's twin imperatives of political control and economic ascension might lead to the twilight of fansub in China. But the demand for diverse content and the therapeutic need to share cannot be easily squashed. The availability of foreign contents with a click will continue to offer fans ample

---

37. Laikwan Pang, *Creativity and Its Discontents China's Creative Industries and Intellectual Property Rights Offenses* (Durham, NC: Duke University Press, 2012), https://doi.org/10.1215/9780822394587.

material for translating and sharing, only if for personal joy, with or without the transgressive aim to subvert established legal and political systems.

Last but not least, when it comes to appropriating/translating contemporary AV content from other languages and cultures, who gets to decide what stories to select and tell, and how? As fansub offers far more lively and relevant interpretation of foreign contents, Chinese audiences have become increasingly averse to the stiff and authoritarian tone of official interpretation, rendering the process of meaning-making via translation a contested arena. Lydia Liu coined the term "translingual practice"[38] in the mid-1990s to encapsulate the role translation played in the dynamic process of meaning-making and culture-building during China's historical encounter with the West (via Japan) during the early Republic era. Such a complex process of linguistic negotiation and interpretive power-jostling came with real world consequences, leading to what Liu calls "translated modernity." Liu's book essentially calls attention to the role of Chinese writers and intellectuals in shaping the course of China's early modernity. How might the politics of translingual practice in the form of vernacular fansubbing play out in contemporary youth's cultural appropriation? Liu argues that the arrival of modern Chinese literature is the result of "a cross-cultural breeding facilitated by translation." And, as such, "what to translate and why" become the key questions. As Liu frames it, "In whose terms, for which linguistic constituency, and in the name of what kinds of knowledge or intellectual authority does one perform acts of translation between cultures?"[39] The ability of grassroots fansub in absorbing and adapting to local tastes is an interesting exercise, to repurpose Liu's phrases in her articulation of an earlier era, in "active transcultural building" or "cultural production through translation," in the popular audiovisual arena.[40] Meanings are thus

---

38. Lydia He Liu, *Translingual Practice Literature, National Culture and Translated Modernity—China, 1900–1937* (Stanford, CA: Stanford University Press, 1999).
39. Liu, *Translingual Practice Literature*, 1.
40. Liu, 26.

not so much "transformed" as invented within the local environment. In this regard, fansub plays a crucial role in the local construction of networked knowledge and perspective. In a nutshell, the foreign audiovisual contents filtered through Chinese fansub have influenced Chinese youth perception and experience of events, past and future, local and global, and helped to shape their cultural affiliations and identifications. Future research might compare the experiences of the laborious "translated modernity" in China's earlier top-down and elite encounter with Western languages and literatures to that of grassroots youth appropriation of Western popular culture in a more effortless and instantaneous fashion.

Japanese Dramas and the Streaming Success Story that Wasn't

How Industry Practices and IP Shape Japan's Access to Global Streaming

DAVID HUMPHREY

Abstract

I examine the relative absence of Japanese dramas in the global streaming landscape. Contrasting this to their 1990s and 2000s boom in East Asia, I argue that structural rather than cultural factors play a central role in this development. At present, a split exists between the domestic market for Japanese dramas and the transnational one, where only a few, largely off-mainstream dramas find traction. Through a discussion of streaming's prehistory and present in Japan, I contend that the split stems, in part, from three issues: television networks' continued dominance, industry practices that favor advertising-based revenue models, and the adoption and impact of international intellectual property regimes on Japan's media industry in general. I argue that these issues and the split that they inform stand to have significant consequences for Japanese dramas, as well as other media genres, in the emergent era of streaming. Although Japan's media industry has largely been able to navigate the incursion of US-based streamers into its domestic market and maintain control there, it has become increasingly reliant on mostly US-based platforms to distribute its content abroad.

Keywords: streaming, intellectual property (IP), industry studies, Japanese dramas, Japanese media

https://doi.org/10.3998/gs.3668

In this article, I examine the relative absence of Japanese dramas on the global scene in the present era of international streaming—an absence that I will posit stems from structural rather than cultural factors. Japan's media industries are, in a sense, split. On one hand, a domestic television-dominated industry continues to serve up dramas for Japan-specific audiences while, on the other, a transnational streaming-oriented one—a system increasingly dominated by US-based platforms like Netflix—distributes more off-mainstream fare whose audience both within and outside Japan lean, more often than not, toward the niche. In interrogating this split, my analysis will focus primarily on the media industry that produces Japanese dramas and the network structures that facilitate their distribution, rather than the dramas themselves. Ultimately, a content-based study of the handful of Japanese dramas that do find their way to streaming outside of Japan would be an underwhelming and rather myopic exercise. Moreover, it would leave unaddressed the structural factors that inform the dearth of Japanese dramas on streaming outside of Japan. The motivating question of this article, therefore, will be why this lack exists to begin with. Responding to this question, my tentative answer will focus on three principal areas: the enduring influence of the television industry in Japan, the advertising-based revenue strategies that inform its practices, and the ascendant role of transnational intellectual property (IP) regimes in the age of streaming.

Recent research highlights television's continued impact on media networks that operate in and around Japan while it also elucidates the struggles, which Japan's media industry faces against that backdrop. In a discussion of the *Hanayori dango* (Boys over flowers) franchise—a franchise that has been adapted across several East Asian media markets—Thomas Lamarre highlights how television's logic, although perhaps no longer the central force it once was, has been integrated within networks of distribution and consumption in East Asia.[1] More broadly, Shinji Oyama provides an overview of the continued power that Japanese television networks and advertising giants like

---

1. Thomas Lamarre, "Regional TV: Affective Media Geographies," *Asiascape: Digital Asia* 2, no. 1–2 (2015): 94–6, 104–13.

Dentsu wield in Japan where they maintain a mutually sustaining relationship with the networks.[2] Similarly, Marc Steinberg brings together the disparate threads of research on streaming in Japan and highlights the still untapped potentials of Japanese streaming platforms, such as ABEMA (previously AbemaTV). These platforms, according to Steinberg, provide glimpses of how streaming services might branch out to integrate more familiar video-on-demand (VOD) services with those that mimic older linear broadcast.[3] More generally, Steinberg's work on digital platforms highlights how IP practices and regimes have informed and continued to inform the ways in which platforms manage content and, moreover, in which they have ushered in an era of what Anne Helmond identifies as "platformization." Describing the manner in which content and other cultural transactions come to conform to the logic of digital platforms, platformization represents a process key to understanding VOD services, which can themselves be classified as platforms.[4]

Alongside this more recent work, earlier research from the 2000s, led by Iwabuchi Kōichi's groundbreaking work on the transnationalization of media in Japan, offers important back-history to understanding the present-day intersection of streaming and Japanese dramas.[5] Although this earlier body of research is necessarily shaped by a prestreaming perspective, it provides a vital foundation for situating how Japanese dramas circulate or fail to circulate within global contexts, particularly East Asian ones. As I will argue throughout, the 1990s and 2000s spread of Japanese dramas, which Iwabuchi and

---

2. Shinji Oyama, "Japanese Creative Industries in Globalization," in *Routledge Handbook of New Media in Asia*, ed. L. Hjorth and O. Khoo (London: Routledge, 2015), 325–27.

3. Marc Steinberg, "AbemaTV: Where Broadcasting and Streaming Collide," in *From Networks to Netflix: A Guide to Changing Channels*, 2nd ed., ed. Derek Johnson (New York: Routledge, 2022), 347–56.

4. Marc Steinberg, *The Platform Economy: How Japan Transformed the Consumer Internet* (Minneapolis: University of Minnesota Press, 2019), 43–45, 49; Anne Helmond, "The Platformization of the Web: Making Web Data Platform Ready," *Social Media + Society* 1, no. 2 (2015): 1–11.

5. Iwabuchi Kōichi, *Toransunashonaru Japan: Ajia o tsunagu popyurā bunka* (Recentering globalization: Popular culture and Japanese nationalism) (Tokyo: Iwanami shoten, 2001).

other scholars of the era studied, foreshadowed both the possibilities and difficulties Japanese dramas would encounter with the transnational spread of VOD platforms. Notably, in her analysis of the Japanese drama boom of the 1990s and 2000s, Gabriella Lukács argues that the piracy that fueled its spread throughout East Asia threw into relief the decentralizing forces that accompanied the centralizing ones of transnational media corporations with the rise of media globalization.[6]

I turn first to a brief examination of Japanese dramas' 1990s and 2000s success throughout East Asia and highlight how the underwhelming fate of Japanese dramas at present was not preordained. Taking up the idea that cultural barriers hampered Japanese dramas success abroad, I contend that the success of Japanese dramas abroad during these years in fact casts doubt on the capacity of the cultural hypothesis to fully explain the difficulties Japanese dramas face. In its place, I consider the impact of the Japanese media industry and state's move in the 2000s to seek stricter IP enforcement—part of a push to establish Japan, as the state bureaucracy then put it as an "IP-oriented nation" (*chiteki zaisan rikkoku*).[7] I propose that such moves and the structural barriers, which they reinforced, more significantly hampered Japanese dramas' circulation abroad with the rise of streaming. I then turn to an examination of the present state of streaming and dramas in Japan to highlight how IP practices and advertising-based revenue models continue to inform a split landscape between domestic and nondomestic media networks. Here, the television networks continue to dominate domestic media in Japan, even as streaming has gained greater currency and begun to serve as a credible platform for dramas there. Indeed, at the same time that US-based streamers, such as Netflix and Amazon Prime, have made inroads

---

6. Gabriella Lukács, *Scripted Affects, Branded Selves: Television, Subjectivity, and Capitalism in 1990s Japan* (Durham, NC: Duke University Press, 2010), 178–79.

7. Chiteki zaisan senryaku honbu (Intellectual Property Strategy Headquarters), *Chiteki zaisan sōzō, hogo oyobi katsuyō ni kansuru suishin keikaku* (Plan regarding the advancement of intellectual property creation, protection, and use), July 8, 2003, https://www.kantei.go.jp/jp/singi/titeki2/kettei/030708f.pdf.

among Japanese viewers over the past decade, the Japanese media industry has developed credible alternatives that more specifically match the dynamics of Japan's domestic media market, such as so-called *minogashi* (catch-up) streaming services. Although this situation has allowed Japanese media companies to maintain control of drama production and circulation within Japan, it has rendered them largely dependent outside of Japan—with a few exceptions regionally in Asia—on non-Japanese, generally US-based streaming platforms. Drawing on Dal Yong Jin's concept of platform imperialism,[8] I highlight in conclusion how this situation extends older forms of cultural imperialism, with a new twist: on its surface, content available to streaming viewers around the world has become increasingly diverse in terms of national origin, but beneath that surface, transnational platforms act as gatekeepers, controlling the procuration and distribution of that content.

From Boom to Bust

Data on the sale and distribution of Japanese dramas overseas underscores how they have languished over the past decade, even as anime has successfully harnessed the expansion of over-the-top (OTT) streaming services worldwide. Annual reports on the export of broadcast content from Japan's Ministry of Internal Affairs and Communications (MIC) show that, although overall exports of Japanese-produced media have posted healthy growth from 2013 to the present, exports of dramas in specific have stagnated. Overseas sales of Japanese dramas—inclusive of sales in broadcast rights, disc resale, Internet streaming, merchandizing, and so forth—increased in the early 2010s from 1.71 billion yen ($12 million USD) in 2012 to 3.11 billion yen ($22 million USD) in 2014. However, these numbers have oscillated in subsequent years, dipping to 2.67 billion yen ($19 million USD) in

---

8. Dal Yong Jin, *Digital Platforms, Imperialism and Political Culture* (London: Routledge, 2015).

2015 before rising again to 3.5 billion yen ($25 million USD) in 2017. By 2020—the last year for which MIC data is available—drama sales had again fallen to 2.54 billion yen ($18 million USD). In contrast, international sales of anime expanded exponentially, going from 5.9 billion yen ($42 million USD) in 2012 to an eye-popping 49.63 billion yen ($360 million) in 2020.[9] As Eva Tsai has highlighted, one hidden avenue of Japanese dramas' overseas distribution is through remakes throughout Asia.[10] However, such data underscores the limits of this route to international access, particularly as streaming comes to dominate. Notably, the MIC's 2022 document (for data to 2020) reports that, while overall revenue from concept licensing for all genres has remained relatively steady since 2013, its percentage of overall revenue from broadcast content exports has fallen in comparison to broadcast rights and streaming.[11]

Such figures might appear commonsensical: the popularity of Japanese anime outside of Japan is at present indisputable, while Japanese dramas remain relatively obscure. However, it was not always this way. In the 1990s and early 2000s, Japanese television dramas—especially those of the so-called trendy

---

9. MIC Institute for Information and Communications Policy, *Hōsō kontentsu no kaigai tenkai ni kansuru genjō bunseki (2013 nendo)* (Analysis of the present state of the development of broadcast contents overseas [2013 edition]), November 28, 2014, https://www.soumu.go.jp/main_content/000324498.pdf; MIC Institute for Information and Communications Policy, *Hōsō kontentsu no kaigai tenkai ni kansuru genjō bunseki (2015 nendo)* (Analysis of the present state of the development of broadcast contents overseas [2015 edition]), April 10, 2017, https://www.soumu.go.jp/main_content/000477810.pdf; MIC Institute for Information and Communications Policy, *Hōsō kontentsu no kaigai tenkai ni kansuru genjō bunseki (2017 nendo)* (Analysis of the present state of the development of broadcast contents overseas [2017 edition]), May 31, 2019, https://www.soumu.go.jp/main_content/000623342.pdf; MIC Institute for Information and Communications Policy, *Hōsō kontentsu no kaigai tenkai ni kansuru genjō bunseki (2020 nendo)* (Analysis of the present state of the development of broadcast contents overseas [2020 edition]), June 3, 2022, https://www.soumu.go.jp/main_content/000817306.pdf.

10. Eva Tsai, "Remade by Inter-Asia: The Transnational Practice and Business of Screen Adaptions Based on Japanese Source Material," in *Routledge Handbook of Japanese Media*, ed. F. Darling-Wolf (New York: Routledge, 2018), 388–402.

11. MIC Institute for Information and Communications Policy, *Hōsō kontentsu no kaigai tenkai ni kansuru genjō bunseki (2020 nendo)*.

drama subgenre—enjoyed popularity throughout East Asia, particularly among youth audiences. Building on the Pan-Asian popularity of 1980s television dramas like the NHK telenovel *Oshin* (1983–84), youth-oriented star vehicles like *Tokyo Love Story* (Fuji TV, 1991), and *Long Vacation* (Fuji TV, 1996) attracted audiences across the region in the 1990s and early 2000s, riding a wave of media-market liberalization and the introduction of satellite television.[12]

Then as now, anime held a lead over dramas, but the gap was less pronounced. In a 2006 article in Fuji Television's *AURA* magazine, Hara Yumiko, a researcher with the NHK Broadcasting Culture Research Center, noted that the per-hour export of Japanese dramas and variety shows had increased in recent years, citing data from a 2001 survey by the International Communication Flow Project-Japan.[13] Similarly, a 2001 article in *Nikkei Business* suggested that ratings for Japanese dramas, like the 1990s series *Hitotsu no yane no shita* (Under one roof, Fuji TV, 1993, 1997), far outstripped those of US series like *Friends*.[14] By all accounts, Japanese-made television dramas appeared poised to continue strong growth, trending upward with, if not alongside, anime. Yet, as anime's international viewership exploded in the 2010s, seemingly catalyzed if not driven by the spread of streaming, Japanese dramas' viewership and associated cachet all but disappeared. Exceptions to this rule do exist: the 2013 workplace drama *Hanzawa Naoki* and its 2020 sequel were widely viewed across East Asia, leading many to greet it as the new *Oshin*, but these exceptions appear to be precisely that—outliers in an otherwise downward turn for Japanese dramas.[15]

---

12. Lukács, *Scripted Affects*, 179–81, 198.
13. Hara Yumiko, "Nihon no terebi bangumi no kokusaisei—2001–2 nen ICFP chōsa kara" (The international aspect of Japanese television programs: From the 2001–2 ICFP survey), *AURA*, December 2016, 9.
14. "'Nihon daisuki' Ajia wakamono—Fashon, dorama . . . poppu bunka ga shintō" (Asian youth who "love Japan": Fashion, dramas . . . pop culture is spreading), *Nikkei Business*, January 15, 2001, 29.
15. "'Oshin' o koeru ka 'Hanzawa Naoki' Taiwan, Honkon de būmu, Ajia zeniki ni" (Has it overcome "Oshin"? "Hanzawa Naoki" booms in Taiwan and Hong Kong and spreads to all areas of Asia), J-CAST nyūsu (J-CAST news), November 1, 2013, https://www.j-cast.com/2013/11/01187976.html.

A common explanation for the gap between anime and drama's divergent fates is the varying degree to which culturally specific knowledge is needed to decode the two genres. Already by the 2000s, observers inside Japan had begun to put forward this thesis to explain what was then a smaller yet noticeable disparity between the overseas popularity of Japanese anime and dramas. Anime, according to this argument, has very few cultural markers, depicting situations and characters that were both culturally and racially ambiguous. Dramas, on the other hand, use human actors and depict more real-life scenarios and by necessity carry more baggage in this regard. Hara—the NHK researcher cited above—concluded in her 2006 discussion that content categories other than anime likely struggled to gain traction outside of Asia due to the greater likelihood that they depict culturally situated scenarios and/or require knowledge of Japan to comprehend. Anime on the other hand, according to Hara, has a *mukokusekisei* (nonnational character).[16] Similarly, in his studies of the transnational reorientation of Japan toward the Asian market in the 1990s and 2000s, Iwabuchi Kōichi frequently made much the same argument, contending that anime succeeded where other forms like dramas did not, because the former was culturally "odourless." Dramas, on the other hand, according to Iwabuchi, had a specific Japanese "odour" or "smell" that "evoke[d] images or ideas of a Japanese lifestyle."[17]

Yet, the 1990s and early 2000s success of Japanese dramas throughout East Asia—and later of Korean dramas globally—suggest the limits of the cultural specificity argument. As Lamarre suggests, Iwabuchi's cultural odor thesis implicitly rests upon an assumption that there exists a set of cultural practices and norms that could be considered more authentically Japanese, an assumption that echoes the essentialism that Iwabuchi himself rejects.[18]

---

16. Hara, "Nihon no terebi bangumi," 10.
17. Koichi Iwabuchi, "Marketing 'Japan': Japanese Cultural Presence under a Global Gaze," *Japanese Studies* 18, no. 2 (1998): 167. Iwabuchi makes much the same point in his 2001 book. Iwabuchi, *Toransunashonaru Japan*, 30–33.
18. Lamarre, "Regional TV," 110–1.

Notably, Iwabuchi observed that the cultural specificity of dramas and similar forms appeared to have aided their spread in the Asian market during the height of their popularity there during the 1990s and 2000s. Iwabuchi reported, for example, that Taiwanese viewers found *Tokyo Love Story* more relatable than a US show like *Beverly Hills 90210*, since the characters' lives appeared more similar to their own.[19]

Although this would appear to suggest that cultural similarities across East Asia propelled Japanese dramas' popularity in ways that would not be possible in other regions, other studies of the Japanese drama boom at the time suggest a more complicated picture. In a study of the cultural allure of Japanese dramas for young audiences in China at the time, Wu Yongmei argued that, while cultural proximity represented part of the story, a bigger part was the attractiveness of the Japanese consumer lifestyle depicted in dramas like *Tokyo Love Story*, as well as others such as *Long Vacation*.[20] Wu's argument makes sense, since, as Lukács has discussed at length, such dramas—representative of the trendy drama genre that saw its heyday during the 1990s and early 2000s—often functioned largely as vehicles for the visualization of lifestyle trends and the sort of aspirational consumption that the characters portrayed.[21] Wu contends that, far from a nationally nonspecific image of present-day global consumerism, this image of conspicuous consumption, which trendy dramas spread as they circulated through Asia, became closely tied to a specific (if largely fictive) conception of modern

---

19. Iwabuchi, *Toransunashonaru Japan*, 225–26

20. Wu Yongmei, "Puchiburu kibun to Nihon no terebi dorama" (The petite bourgeois feeling and Japan's television dramas), in *<I> no bunka to <jō> no bunka: Chūgoku ni okeru Nihon kenkyū* (The culture of <ideas> and the culture of <feelings>: Japanese studies in China), ed. Wang Min (Tokyo: Chūōkōron shinsha, 2004), 23–27.

21. Notably, in her discussion of the Japanese trendy drama popularity during this period abroad, Lukács writes, in echoes of Iwabuchi, that she is "skeptical" that consumption of the dramas reflected an interest in Japan per se. However, Lukács's skepticism is not necessarily at odds with Wu's analysis and draws instead a contrast between the consumer lifestyle featured in the dramas and a specific interest in Japan as a nationally bounded cultural entity. Lukács, *Scripted Affects*, 40–45, 197–98.

Japanese lifestyles within the minds of the dramas' Pan-Asian audiences, particularly those in mainland China.[22]

These circumstances—alongside the divergent fate of anime with the transition to streaming internationally—suggest that more fundamental, structural issues have since hampered the ability of Japanese dramas to replicate their earlier success. Indeed, in his analysis of streaming and the Japanese media industry, the media journalist and observer Nishida Munechika highlights the role IP practices have played in the varying fates of Japanese anime and dramas in the age of streaming. Nishida notes that, although anime in Japan may be subdivided into various subgenres based on target audience and business model, the large proportion of anime being exported outside of Japan, on streaming or otherwise, derives from anime that, marketed to older audiences, has long been broadcast on television during the late-late-night hours. Historically, broadcasters have perceived these time slots to be much less lucrative in terms of advertising revenue, since viewers who consumed the content broadcast at these hours recorded the programming for later, ad-free viewing. For this reason, broadcasters and anime producers developed a business model different from that of daytime television. In contrast to the latter in which advertising drives broadcasters' revenue, producers of late-night anime, according to Nishida, pay to have their content broadcast during the time slots, with the rationale having long been that the anime broadcast served as the advertisement itself (i.e., as an advertisement for disc sales and other tie-in products in what is elsewhere known as the media mix). (Note: Nishida does not use this specific term.)[23]

---

22. Wu, "Puchiburu," 26. See also Nakano Yoshiko and Wu Yongmei, "Puchiburu no kurashi-kata: Chūgoku no daigakusei ga mita Nihon no dorama" (The petite bourgeois lifestyle: Japanese dramas as seen by Chinese university students), in *Gurōbaru purizumu: <Ajian dorimu> toshite no Nihon no terebi dorama* (Global prism: Japanese television dramas as the <Asian dream>), ed. Iwabuchi Kōichi (Tokyo: Heibonsha, 2003), 185–86, 204–7.

23. Nishida Munechika, *Nettofulikkusu no jidai: Haishin to sumaho ga terebi o kaeru* (The age of Netflix: Streaming and smartphones change television) (Tokyo: Kōdansha gendai shinsho, 2015), 87–88. For a discussion of media mix, cf. Marc Steinberg, *Anime's Media Mix: Franchising Toys and Characters in Japan* (Minneapolis: University of Minnesota Press, 2012).

Nishida contends that this peculiarity of anime production and broadcast positioned it to make a smooth transition to streaming, whereas the more typical practices of the television industry have hampered daytime and evening content like dramas. Notoriously, dramas and other domestically popular television content in Japan, such as variety shows, are governed by a so-called bundled rights model (in Japanese, *kenri no taba*), in which a single broadcast can incorporate a disparate collection of rights holders, from the networks that may hold the broadcast rights for the program to performers and their agencies who might hold the copyrights for songs used and the performers' individual portraiture rights.[24] As Nishida notes, these multiple, bundled rights present a complicated barrier to redistributing Japanese broadcast content over streaming, since it takes only one rights-holder to stymy a deal for redistribution over streaming or other formats. (Nishida likewise notes the stark contrast this model presents vis-à-vis that which Netflix has established for its original content, by which it secures worldwide rights before production.) For content like dramas that use live actors, these difficulties are multiplied tenfold, as rights clearance must be received from the performer (or more typically their agency). Due to its lack of human performers, anime has in comparison little overhead in terms of rights management and clearance.[25]

Nishida argues that anime's roots in late night television furthermore fostered within the industry an early-adopter mindset. As early as the 2000s, for example, anime franchises such as Gundam had begun experimenting with postbroadcast Internet streaming of episodes. According to Nishida, this early embrace of streaming, pursued in spite of the fact that it would potentially divert viewers from the broadcast, stemmed from the fact that anime producers perceived the initial broadcast, as noted above, as

---

24. "Hōsō bangumi ni kansuru kenri shori" (The treatment of rights related to broadcast programs), Nihon minkan hōsō renmei (The Japanese Commercial Broadcasters Association), accessed November 21, 2022, https://j-ba.or.jp/category/minpo/jba101970.
25. Nishida, *Nettofulikkusu no jidai*, 88–89.

an advertisement for other revenue-creating components of a given show's media franchise. Moreover, the anime producers had greater freedom in deciding to offer streaming, since the unique set up of the late-late-night broadcasts allowed them to main greater control over the programs' associated rights.[26]

The relative success of Korean dramas provides an additional point of contrast, as they have similarly seen more sustained growth in terms of international audience from the 2000s onward. In their study of the globalization of the Hallyu wave, Dal Yong Jin, Kyong Yoon, and Wonjung Min report that interviewees from Latin American countries expressed less interest in dramas than in other aspects of the Korean boom, such as music, and thus they infer that cultural contextual issues create a higher barrier to international acceptance for dramas. Here, they conjecture that the often more "old-fashioned" feel of the drama clashes with the image of "hypermodernity" that fans associate with the Korean wave.[27] Despite these challenges, Korean dramas appear to be doing much better than their Japanese peers on a transnational scale: a 2015 MIC white paper reports that Korea exported a total of $309 million worth of broadcast content in 2013, 92.3 percent of which ($208.6 million) came from the sale of broadcast rights. Dramas furthermore represented the overwhelming share of broadcast rights sales, at 88.3 percent. In contrast, Japan saw $104.1 million worth of broadcast exports in the same year, $64 million or 45.1 percent of which was for broadcast rights. Of this sum, dramas' share of the pie was a mere 18.1 percent.[28]

26. Nishida, 83–85.

27. Dal Yong Jin et al., *Transnational Hallyu: The Globalization of Korean Digital and Popular Culture* (New York: Rowman & Littlefield, 2021), 99–100.

28. MIC Institute for Information and Communications Policy, *[Sankō] Hōsō kontentsu no kaigai tenkai ni kansuru kokusai hikaku (2013 nen)* ([Reference] International comparison of the overseas development of broadcast contents [2013 edition]), March 2015, https://www.soumu.go.jp/iicp/chousakenkyu/data/research/survey/telecom/2014/broadcasting-contents-ex2013-ref.pdf.

Much of these drama exports appear to have been to other Asian markets, as the bulk of Korean content exports according to the MIC report went
to Japan and other Asian countries. Yet, recent reports indicate that Korean
dramas have begun expanding their appeal to non-Asian markets, suggesting
that the perceived cultural barriers are not in fact insurmountable and that
other media exports such as the more familiar K-pop provide a bootstrapping
effect.[29] Indeed, Jin et al. conclude much the same, hinting at the role that
digital media play. They furthermore note that the Korean wave's success has
partly been a consequence of the Korean industry's embrace of digital media,
from social media platforms like Facebook and YouTube to streaming ones
like Netflix, where media convergence allows disparate content forms to pro-
pel each other's transnational circulation.[30] It is moreover worth noting that
this embrace of digital platforms by the Korean entertainment industry has
often been attributed to a more lax attitude toward copyright infringement,
which allowed for a more flexible approach to the sharing of Korean popular
culture on the services.[31] This contrasts the situation in Japan's entertainment
industry where influential talent agencies often take a much more conserva-
tive approach toward controlling the circulation of their celebrities' images
on the Internet, typically through the use of portraiture rights. In a telling
example, the idol agency Johnny & Associates—an agency notorious for its
aggressive use of portraiture rights—only allowed its male idol group Arashi
to create public accounts on Twitter, Instagram, Weibo, and other social
media sites in 2019, the year of the group's twentieth anniversary.[32]

29. For example, Sara Layne and Cynthia Littleton, "K Dramas Can't Be Denied: Global Stream-
    ing Spurs Demand for Asian Content Platforms," Variety, August 18, 2022, https://variety.
    com/2022/streaming/news/korean-dramas-kocowa-viki-asiancrush-kcon-1235344275/.
30. Jin et al., Transnational Hallyu, 7–8, 149–50.
31. Shin Dong Kim and Jimmyn Parc, "The Digital Transformation of the Korean Music
    Industry and the Global Emergence of K-Pop," Sustainability 12, no. 18 (2020), https://
    doi.org/10.3390/su12187790.
32. "Arashi no netto kaikin, Janīzu no ijōsei ga ukibori ni . . . SMAP kaisanji no akumu
    soshi ka" (With the raising of the net embargo for Arashi, the abnormality of Johnny &
    Associates is thrown into relief . . . A preemption of the nightmare of SMAP's break up?),
    Business Journal, November 5, 2019, https://biz-journal.jp/2019/11/post_126531.html.

Japanese dramas' failure to see sustained growth on par with Korean dramas following their early 2000s peak arguably confirms Nishida's thesis that IP restrictions have presented one of the greatest hurdles for Japanese dramas in the era of international streaming. Ironically, like the story of Japanese dramas themselves, the Japanese industry's comparatively stringent approach to copyright enforcement appears to be a relatively recent development. As Kelly Hu documents, the boom in Japanese dramas across East Asia in the 1990s and the early 2000s was fueled in large part by the circulation of pirated video CD (VCD) copies of popular dramas.[33] During the 1990s, satellite broadcast had provided the first beachhead for Japanese dramas and other broadcast content abroad, particularly in Taiwan where they gained popularity following the 1993 relaxation of broadcast restrictions on Japanese-language media that had been in place throughout the postwar era.[34] However, fans of Japanese dramas in Taiwan and other East Asian countries had to wait years for many of the programs to broadcast locally, leading them to turn to pirated VCDs originating largely in Taiwan and Hong Kong. The VCD format boomed in East Asia during the 1990s and 2000s, particularly among young people who could play them on either dedicated players or their own computers. The VCD piracy networks furthermore laid the groundwork for Internet-based peer-to-peer (P2P) piracy later in the 2000s with the spread of reliable broadband; they also functioned as a major conduit for the more generalized "Japan boom" of the 1990s and 2000s, characterized by an interest in Japanese music, fashion, and lifestyle throughout East Asia.[35]

Curiously, Japanese media companies assumed at the time what might be described as a relatively hands-off attitude toward piracy. While Hong Kong sought to crack down on VCD piracy in the late 1990s, this effort

---

33. Kelly Hu, "Chinese Re-makings of Pirated VCDs of Japanese TV Dramas," in *Feeling Asian Modernities: Transnational Consumption of Japanese TV Dramas*, ed. Koichi Iwabuchi (Hong Kong: Hong Kong University Press, 2004), 215–16.
34. Iwabuchi, "Marketing 'Japan,'" 173–76; Hu, "Chinese Re-makings," 212.
35. Wu, "Puchiburu," 33–40.

was initiated by Hong Kong authorities and apparently not at the request of Japanese media companies. Reportedly, the former had to approach the latter first in order to pursue the crackdown.[36] During these years, the Japanese entertainment industry took what Hu characterizes as a "passive" approach to the rampant piracy and showed little interest in preventing it.[37] Hu argues, however, that rather than an openness to piracy, this laissez-faire attitude appeared driven more by a lack of interest in the East Asian market and the belief that leveraging drama sales there would not prove profitable.[38]

This hands-off posture shifted, however, in the early to mid-2000s as Japanese media companies began to awaken to the lucrative possibilities of Japanese dramas' popularity with youth audiences in East Asia, and with this new awareness, those same media companies, as well as the Japanese government, began to push for a clampdown on copyright infringements across East Asia.[39] In addition to a growing awareness of piracy, the expanding reach of the Internet and spread of broadband loomed large in such moves. With the rollout of digital broadcasting in the 2000s, for example, television networks pushed for the inclusion of anticopying technologies in the new standards, likewise out of concerns that digital recordings of broadcast were making their way abroad via the Internet.[40] Similarly, in a 2003 report on its nascent intellectual property strategy, the prime minister's office of Japan declared its intent to establish Japan as an "IP-oriented nation" (*chiteki zaisan rikkoku*). Under this policy, it proposed to pursue greater piracy enforcement domestically and advocate for measures targeting foreign markets. In

---

36. Darrell William Davis and Emilie Yueh-yu Yeh, "VCD as Programmatic Technology: Japanese Television Drama in Hong Kong," in *Feeling Asian Modernities: Transnational Consumption of Japanese TV Dramas*, ed. Koichi Iwabuchi (Hong Kong: Hong Kong University Press, 2004), 233–34.

37. Hu, "Chinese Re-makings," 216–17.

38. Hu, 215, 217.

39. Hu, 218; Yamada Shōji, "Bunka kakusan to chizai hogo no sōkoku" (Conflicts between the cultural diffusion and the intellectual properties protection), *IPSJ SIG Technical Reports*, March 19, 2005, 3.

40. David Humphrey, "The Black Box and Japanese Discourses of the Digital," *International Journal of Communication* 14 (2020): 2486–87.

justifying the move, the report cited estimates that Japanese content piracy in mainland China led to yearly losses of around 2 trillion yen ($15.3 billion USD), suggesting that the smoother flow of data made possible by the Internet played no small part.[41]

While it is unwise to draw a direct cause-effect relationship—and certainly beyond the scope of this article to effectively do so—it is nonetheless striking that this clampdown coincides with the general time frame (i.e., the latter half of the 2000s) when Japanese drama's popularity outside Japan appears to have peaked and eventually declined. Given this coincidence of the two time lines, it is possible to surmise that the awakening to international piracy and the turn in the early 2000s to greater IP enforcement lay the stage in some respect for Japanese dramas' transnational decline. Indeed, as I will demonstrate in the following section, Japanese television networks leveraged their control of the rights to popular domestic content like dramas as part of a strategy to slow-walk the transition to streaming in Japan. Although this ostensibly benefited the networks in the short term, allowing them to maintain their dominance of the domestic media landscape, it has had questionable results in the long term. Notably, the efficacy of the turn to IP enforcement has been questionable at best. As a 2022 document from Japan's Agency for Cultural Affairs declared with no little consternation, piracy of Japanese cultural contents in Asia—China in particular—continued and perhaps even expanded during the 2010s.[42] More pressingly, perhaps, the move to double down on IP restrictions appears to have played one part in dampening the networks' ability to circumvent US-based streamers' dominance globally, making them largely dependent on the platforms for distribution outside of Japan.

---

41. Chiteki zaisan senryaku honbu, *Chiteki zaisan sōzō*.
42. Agency for Cultural Affairs, Government of Japan, *Intānetto jō no chosakuken shingai (kaizokuban) taisaku handobukku—Chūgokuhen* (Handbook of strategies for addressing copyright infringement on the internet [piracy edition]: China edition), March 2022, 1–6, https://www.bunka.go.jp/seisaku/chosakuken/kaizoku/assets/pdf/kaizokuban_hand book_chn.pdf.

Streaming's Uneven Adoption and Japanese Dramas' Present

At present, the OTT streaming market remains comparatively small in Japan. A 2022 report from the Tokyo-based marketing and research firm GEM Partners estimated the total value of the VOD market in Japan in 2021 to have reached 461.4 billion yen ($3.3 billion USD), fueled in part by the turn to streaming during the COVID-19 pandemic.[43] However, use of (and even awareness of) streaming in comparison to other media and platform types appears to remain relatively low. A survey by the Softbank-owned news site ITmedia revealed that, although 25.6 percent of respondents reported in 2021 to having used a streaming service in the past three months (up from 7.7 percent in 2015), an overwhelming majority continued to favor real-time television broadcast, with 69.9 percent of respondents reporting that they often viewed linear broadcasts and only 30.7 percent and 25.6 percent reporting that they regularly use free and paid streaming services, respectively. Video-sharing services, such as YouTube, enjoyed more robust numbers, with 45.9 percent of respondents reporting using them frequently. Furthermore, when broken down by age and gender, video-sharing sites topped or pulled mostly even with linear broadcast among teen and twenty-something male and female demographics, as well as thirty-something males.[44]

---

43. GEM Standard, "<Dōga haishin (VOD) shijō kibo> 2021 nen VOD shijō zentai wa zen'nenhi 19.0% zō no 4,614 oku en, SVOD shijō shea de 'Netflix' 3 nen renzoku No. 1, 'Dizunī purasu' yakushin" (<The scope of the video streaming (VOD) market> The total 2021 VOD market increased 19.0% over the past year to 461.4 billion yen, "Netflix" remained No. 1 in terms of SVOD market share for 3rd year in a row, "Disney Plus" jumped ahead), press release, February 18, 2022, https://gem-standard.com/statics/download/Press_Release_VOD_Market_in_Japan_2021_Ja.pdf.

44. Inpuresu sōgō kenkyūjo, "Yūryō no dōga haishin sābisu riyōritsu wa 25.6%, koronaka de dōga shichō sutairu ga gekihen 'Dōga haishin bijinesu chōsa hōkokusho 2021'" (Fee-based video streaming service usage 25.6%, viewing styles changed radically with spread of coronavirus "2021 video streaming business survey report"), press release, May 20, 2021, https://research.impress.co.jp/topics/list/video/625.

Within the OTT market, streaming options are primarily split between foreign-based operations and a variety of homegrown services. Most analyses position the US-based Netflix and Amazon Prime Video as market leaders, followed by fully or partially Japanese-run services such as the Japan-based U-Next—a service that offers subscribers access to video alongside manga, magazines, and other text media—as well as Hulu, which in Japan is owned and run by a holdings company led by the television network Nippon TV (NTV). The marketing research firm GEM Partners reports, for example, that Netflix and Amazon Prime commanded the largest share of the Japanese subscription video-on-demand (SVOD) market in 2021 (23.1 percent and 12 percent, respectively), with Disney+ rising quickly through the ranks, going from 3.9 percent market share in 2020 to 6 percent in 2021. U-Next and Hulu Japan, on the other hand, had market shares of 11.5 percent and 8 percent, respectively.[45] Other estimates similarly place Netflix and Amazon at the top but differ on the specifics. In a marked contrast, the ITmedia survey ranked Amazon as the hands-down leader in terms of reported use (69.2 percent), followed by Netflix in a distant second (21.4 percent). Hulu Japan and U-Next followed at 10.3 percent and 6.3 percent, respectively.[46]

As these discrepancies suggest, such numbers are not completely reliable. It is difficult to obtain and compare actual usage statistics, as streaming companies do not generally make the data publicly available. Reports like that of GEM Partners base their SVOD market-share ranking on total subscription fees collected by each service, which necessarily favors more expensive services like Netflix. Polls, like ITmedia, base their rankings on reported usage and thus generally place Amazon first, a reversal that can generally be explained by the lower cost of Prime Video, which is generally included within the broader Amazon Prime service.[47] Similarly, other factors can

---

45. GEM Standard, "<Dōga haishin (VOD) shijō kibo>."
46. Inpuresu sōgō kenkyūjo, "Yūryō no dōga haishin."
47. GEM Standard, "<Dōga haishin (VOD) shijō kibo>." Steinberg makes a similar point on streaming data in Japan. Steinberg, "AbemaTV," 349.

skew results. The NTT Docomo-run dTV forerunner dVideo, for example, gained attention in the early 2010s when it appeared to experience an exponential increase in subscribers over a period of a few short years. However, it later came to light that Docomo had been pushing customers to bundle the service with smartphone contracts, leading to inflated subscription numbers that did not reflect actual usage.[48]

These discrepancies in data aside, it is notable that, with the exception of U-Next, the majority of domestically based streaming services are associated with or primarily owned by a television network. Mirroring NTV's majority stake in Hulu Japan, the other major Tokyo-based television networks maintain a range of SVOD, transactional video-on-demand (TVOD), and/or free streaming services. These include Tokyo Broadcasting System's (TBS) Paravi and TV Asahi's TELEASA, both of which are SVOD services, as well as Fuji TV's FOD (Fuji TV on Demand), which offers both a free advertising-based video-on-demand (AVOD) and "premium" fee-based SVOD services. In addition to its homegrown TELEASA, TV Asahi also holds a stake alongside the media and Internet advertising company CyberAgent in ABEMA, which offers both AVOD and fee-based premium SVOD services.[49] The Tokyo-based and regional networks furthermore collaborate on the popular TVer service, a so-called *minogashi haishin* (catch-up streaming) AVOD service that allows users to stream recent episodes of broadcast programs.

This involvement of Japan's television networks in many of the available VOD services—as well as the slower adoption of streaming overall—reflect the continued influence they exercise on Japan's media landscape. Building on Oyama's discussion of developments during the 2000s, Steinberg suggests that the terrestrial broadcasters maintained their dominance by effectively sidelining then-emergent rivals—in particular the Internet start-up Livedoor—which sought to leverage the growth of the Internet and circumvent the barriers to the media market maintained by the legacy companies.

48. Nishida, *Nettofulikkusu no jidai*, 50.
49. Steinberg, "AbemaTV," 350–51.

In the 2000s, Livedoor, run by Horie Takafumi, launched a hostile takeover of Fuji. The bid ultimately failed, and following a subsequent securities fraud investigation, Horie was jailed. Oyama concludes that the incident served as a warning shot to other Internet and media start-ups who would seek to upend television networks' dominance.[50] Steinberg argues that this likewise allowed the networks to stymy streaming during the early 2000s, as they had little incentive to embrace it, and insurgent companies had been forewarned not to challenge the networks' stranglehold on the media market.[51]

In addition to Oyama's and Steinberg's insights, one can add the control over production and rights that networks leveraged in their move to slow-walk the adoption of streaming. Nishida highlights how, although the Japanese press greeted the mid-2010s arrival of US streamers like Hulu and Netflix as an unstoppable and disruptive force equivalent to the arrival of Commodore Perry's *kurofune* (black ships) in the mid-nineteenth century, reality forced the foreign companies to take a much more cautious approach and build partnerships with the domestic networks rather than overtly antagonize them. Nishida argues that the struggles of Hulu's Japanese subsidiary, prior to NTV's 2014 purchase of it, foregrounded the need for such an approach. When the service first arrived in Japan in 2012, subscriptions saw a healthy increase, as its US-dominated content library drew fans of foreign movies and dramas. However, subscriptions soon stagnated, underscoring the limited appeal of such content to broader audiences in Japan. NTV's acquisition of Hulu Japan reversed this trend, since NTV brought offerings from its own broadcast-based library as well as the capacity to produce original Japanese programming for the service. For NTV, the deal gave it access to a preexisting platform and thus allowed it to avoid having to create their own from scratch.[52]

Perhaps the greatest sign of the networks' continued dominance is the success of the minogashi service TVer. In founding the service in 2015,

50. Oyama, "Japanese creative industries," 328–30.
51. Steinberg, "AbemaTV," 348.
52. Nishida, *Nettofulikkusu no jidai*, 55–61.

Japan's major networks and advertising agencies sought most immediately to counteract the spread of its programming on video-sharing sites like YouTube; however, more long term, they aimed to supplement traditional over-the-air linear broadcast and the advertising-based model that supported it, rather than simply replace it.[53] In contrast to the familiar streaming-centric approach of services like Hulu and Netflix, the minogashi service—a term taken from the verb *minogasu,* meaning "to overlook" or "to miss"—offers streaming as a complement to traditional broadcast. In this sense, it resembles earlier network-based streamers outside of Japan, such as HBO Go, which provided existent subscribers a means to view content in a time-lapsed, on-demand manner. However, besides the most obvious difference that it is an ad-based service free to any user accessing it from a Japan-based IP address, TVer differs in how it is presented to viewers. TVer, as it is imagined by the networks that support it, provides viewers, who might have *missed* one or more broadcasts of a show (ergo the term minogashi), a way to catch up on past episodes and thus resynchronize themselves in a sense with the regular broadcast. Nishida proposes that it was in fact the minogashi model that finally broke down television networks' resistance to streaming and allowed for its more widespread adoption in Japan from the mid-2010s onward. Indeed, as viewership for broadcast flagged in the 2010s, data suggested that shows that offered postbroadcast minogashi streaming options saw more sustained ratings from week to week. Networks thus came to see minogashi options like TVer as a means to prop up and continue their existent advertising-based business model.[54]

TVer is not the only service to provide minogashi streaming—many of the other network-based streamers provide similar streaming, either in an

53. Nishida, 70–71; Honda Masakazu, "Bangumi netto haishin 'TVer' wa dōshō imu: Kōkoku dairi ten no aseri ga unda bijinesu moderu" (Program streaming service "TVer" shares the same bed but a different dream: The business model born from advertising companies' panic), Tōyō keizai Online, July 18, 2015, https://toyokeizai.net/articles/-/77520.
54. Nishida, *Nettofurikkusu no jidai,* 91–94.

AVOD or SVOD capacity—but it is perhaps the most popular, presumably due to its ability to serve as a one-stop location for content from across the various networks. According to the ITmedia survey, TVer ranked fifth among free online video sources but first among VOD services in this category. (The first 4 spots were occupied by YouTube's video-sharing service followed by Twitter, the Japan-based social media platform LINE, and Instagram.) TVer and its AVOD peers furthermore enjoy favor among drama viewers, which is perhaps not surprising considering the serial nature of many dramas and the intent of the platforms as catch-up services: 45.8 percent of respondents reported using nonsharing-based VOD services to watch Japanese dramas, followed by 35.9 percent of respondents for variety shows and 31.1 percent for anime. A comparable 39.4 percent of respondents reported using SVOD services to watch Japanese dramas. However, this trailed behind the popularity of other content, with 55.3 percent of respondents reported using SVOD to watch foreign films, 52.7 percent Japanese films; 44 percent anime, 26.1percent non-Korean foreign dramas, and 14.2 percent Korean dramas. Free video-sharing services, like YouTube, appeared to be more likely used for shorter and even nonvideo content, with 48.8 percent reporting using it to listen to music, followed by content such as hobby, cooking, and "Let's Play" (gēmu jikkyō) gameplay videos.[55]

The result of this network-centric streaming ecosystem is a drama production and distribution landscape that sustains rather than replaces the television networks. Naturally, television survives here in a large sense in air quotes, since this television does not strictly mirror the terrestrial broadcast of yesteryear. Instead, it is television as a concept and business model that persists in the abstract while distribution now takes place in a much more decentered fashion and over a more asynchronous timeframe, with viewers watching drama content just as likely on a smartphone as on the regular broadcast. However, it means that many dramas—particularly those that

---

55. Inpuresu sōgō kenkyūjo, "Yūryō no dōga haishin."

attract a broad viewership—are still produced in a manner that is complimentary to the structural imperatives of television and television networks, continuing until recently to enter circulation first with linear broadcast.

It is likely for this reason that few direct-to-streaming dramas have gained traction in the manner seen outside of Japan. This latter point is of course difficult to gauge for the same reasons it is difficult to determine the penetration and adoption of streaming itself. It is nonetheless striking that the Japan-based SVOD services—even the nonnetwork affiliated U-Next—tend to promote broadcast-based dramas prominently on their streaming pages, to which they subordinate their original content, which appears in less prominent places.[56] This appears to reflect that broadcast-based dramas continue to dominate more generally. Kadokawa's quarterly the Television Drama Academy Awards, for example, consistently bequeaths its best drama award to dramas originating from traditional broadcast—a somewhat circular point perhaps, given that the award is structured around broadcast dramas exclusively, but telling nonetheless insofar that the organizers apparently feel no pressure to open the award to streaming-first dramas or even create a category for streaming dramas.[57]

Yet, while the continued dominance of broadcast-based content props up a continuation of the preexisting TV and advertising-centric system, it hampers these media industries' ability to distribute content abroad via streaming. Indeed, the failure of many Japanese dramas to gain access to audiences outside of Japan via streaming appears to largely stem from the self-imposed insularity of a domestic-facing media industry. It is hard to avoid the conclusion that this insularity is due in no small part to that industry's approach to IP, since the export of mainstream dramas to non-Japanese streaming presumably faces numerous hurdles due to the rights-bundling issues described above.

---

56. Based on review of SVOD service pages in fall 2022 by the author.
57. "Jushōreki—Saiyūshū sakuhin shō" (Award history: Outstanding work award), Za terebijon dorama akademii shō (The Television Drama Academy Awards), accessed November 23, 2022, https://thetv.jp/feature/drama-academy/archives/department/best-drama/.

Supporting this conclusion, the dramas that do make it abroad appear to either have been produced specifically for streaming or otherwise smaller productions with likely fewer barriers posed by bundled-rights permissions. To cite one example, the eight-episode *Miss Sherlock*, which won the Asian Academy Creative Awards 2018 award for Best Drama Series, was coproduced by Hulu Japan and HBO Asia, clearing its path to be available on HBO's various national and regional subsidiaries worldwide.[58] Netflix has similarly produced original Japanese dramas that it then makes available outside of Japan as well, such as its 2016 *Hibana: Spark,* based on the Matayoshi Naoki novel of the same name, the 2019 and 2021 seasons of *Zenra kantoku* (The naked director), and, more recently, the 2022 *Shinbunkisha* (The journalist) based on a book by the journalist Mochizuki Isoko. The company has also revived older network-based series with newer streaming-oriented content, as it did with its 2016 and 2019 *Shinya shokudō: TOKYO STORIES* (Midnight diner: Tokyo stories) spin-off of the manga-based *Shinya shokudō* series that aired on Japanese television for three seasons in 2009, 2011, and 2014.[59]

Although many of these dramas enjoy strong followings both domestically and abroad, it is nonetheless striking that they are different in genre and overall production style from the more broadly viewed dramas that first circulate during regular broadcast slots in Japan. Winners of the Television Drama Academy Awards continue to hail from the well-tread genres of family, medical, crime, and school dramas, while these broadcast-first dramas tend to cast familiar stars or at least frequently include high-profile idols and

58. "2018 Final Winners," Asian Academy Creative Awards, accessed November 23, 2022, https://www.asianacademycreativeawards.com/award-ceremony/2018-final-winners/; "Takeuchi Yūko x Kanjiya Shihori 'Misu Shārokku' ga 'AAA's' de saiyūshū sakuhin shō ni!" (Takeuchi Yūko and Kanjiya Shihori's "Miss Sherlock" takes Best Drama Series award at AAAs), *WEB za terebijon* (WEB the television), December 10, 2018, https://thetv.jp/news/detail/172198/.
59. Yamazaki Nobuko, "Futatabi Netflix de haishin! Kobayashi Kaoru wa 'Shinya shokudō' no hirogari o dō mite kita no ka?" (Streaming on Netflix a second time! How does Kobayashi Kaoru view the spread of "Midnight Diner"?), Movie Walker Press, November 2, 2019, https://moviewalker.jp/news/article/210783/.

other celebrities. This contrasts niche-focused and often off-beat dramas, which cast either less familiar performers or, as in the case of *Shinya shokudō*'s Kobayashi Kaoru, performers more readily known for their acting bona fides rather than the sort of media celebrities cast in mainstream dramas to attract viewers. (Notably, this latter type of star often tends to hail from idol agencies like Johnny & Associates, which, as noted above, tend to be overly protective rather than permissive in their approach to talent portraiture rights.) *Shinya shokudō* in fact offers an illuminating point of contrast, straddling as it does the worlds of broadcast and streaming. It is furthermore notable that the drama's broadcast origins lay in the same late-night programming slots that many anime offerings occupy. This is not altogether surprising, however, as such off-mainstream fare has traditionally filled late-night schedules in Japan alongside anime. It might also explain in part the greater ease with which such content has apparently moved to streaming given that, like late-night anime, it might also face lower IP hurdles, but this latter point is unclear. Endō Hitoshi, who has been involved in *Shinya shokudō*'s production over several of its iterations, has noted that the show's producers faced resistance from its original broadcaster, Mainichi Broadcasting System (MBS), when they sought to adapt the program for Netflix.[60]

As streaming is still relatively new, much of this might change in coming years, but it is an open question whether Japanese media companies will be able to maintain their control over even domestic media content. Recent developments suggest that the networks have begun to accept the need to cooperate with streaming partners outside of Japan, but the options available to them appear largely dominated by US platforms. Exceptions to this rule do exist. Since 2016, for example, Fuji TV has licensed its dramas and other broadcast content in Taiwan to the streamer KKTV, a Taiwanese subsidiary of

---

60. "Netflix to terebi 'seisaku gawa ga keiken shita' kettei teki na sa: 'Shinya shokudō' o tōshi eizō sakuhin no kongo o kangaete mita" (The decisive difference between Netflix and television that "the production side experienced": I considered the future of moving image works through the lens of "Midnight Diner"), Tōyō keizai Online, December 18, 2021, https://toyokeizai.net/articles/-/476838.

the Japanese telecom KDDI since 2010.[61] KKTV has since become a venue for Japanese dramas in the country and most recently offered a simultaneous streaming of the 2022 installment of NHK's long-running *Taiga* historical drama series, *Kamakura dono no 13 nin* (The 13 lords of Kamakura).[62]

Despite such regional examples, US platforms appear to be emerging as the dominant pathway for Japanese dramas to reach broader, non-Japanese audiences. Exemplary here are agreements, which TBS has recently entered into with US-based streamers. In 2021, the television network granted Netflix worldwide streaming rights for the most recent remake of the *Japan Sinks* franchise—a ten-part miniseries entitled *Nihon chinbotsu: Kibō no hito* (Japan sinks: People of hope)—allowing Netflix to carry episodes a mere three hours after their original broadcast.[63] In a similar move, TBS and Disney+ announced an agreement in 2021 to make star-vehicle dramas like the ER drama *Tokyo MER* (2021) available on Disney+. As part of the agreement, these dramas are now available in countries where Disney+'s STAR content hub, in which the content is included, is available.[64] This move by

---

61. Fuji Television Network, "Fuji terebi sakuhin ga Taiwan haishin shijo ni mo jōriku! KK-BOX Gurūpu to Taiwan de no dōga kontentsu de senryaku teikei" (Fuji TV programs arrive on the Taiwan streaming market, too! A strategic alliance with KKBOX Group for video contents in Taiwan), press release, July 6, 2016, https://www.fujitv.co.jp/company/news/160706.html; KDDI, "Maruchidebaisu muke ongaku kontentsu haishin gaisha KKBOX Inc. no kabushiki shutoku nitsuite" (Regarding the stock acquisition of KKBOX Inc., a company offering multidevice oriented music contents streaming), press release, December 15, 2010, https://www.kddi.com/corporate/news_release/2010/1215/.

62. Chen Sumi, "Taiwan de moriagaru Taiga Dorama būmu!—'Kamakura dono no 13 nin' no Nittai dōji haishin ga shichō no arikata o kaeta" (Taiga drama boom swells in Taiwan! Simultaneous streaming of "The 13 lords of Kamakura" in Japan and Taiwan changed the shape of the market), Nippon, February 2, 2023, https://www.nippon.com/ja/japan-topics/g02252/.

63. Inoue Masaya, "TBS no 'Netflix dokusen haishin' ni sukeru terebikyoku no yūutsu" (The despondency of the television networks is apparent in TBS's "Netflix exclusive streaming" agreement), Tōyō keizai Online, November 27, 2021, https://toyokeizai.net/articles/-/469960.

64. "TBS and Disney Conclude Streaming Agreement TOKYO MER Coming to STAR on Oct 27—First Japanese Drama to Stream Worldwide on Disney+," TBS Program Catalog, September 29, 2021, https://www.tbscontents.com/news/2001.

TBS accompanies other shifts that suggest that the networks are gradually coming to accept and adapt to streaming. Japan's main networks recently agreed, for example, to allow dramas, beginning in spring 2022, to stream simultaneously on TVer. This ended previous policy that required TVer to stream episodes only after broadcast.[65] Such moves have likely been spurred by the success of TVer in contrast to an otherwise grim outlook for the networks: reportedly, TVer and similar AVOD services saw an increase in advertising revenue in the first quarter of 2022, even as traditional broadcast continued to lose both viewers and advertising revenue.[66]

Whether these shifts will mark a true sea change for Japanese dramas and their access to audiences outside of Japan remains to be seen. Either way, the agreements with US-based streamers in particular highlight how Japanese companies have become increasingly dependent, now more than ever, on non-Japanese companies and their infrastructure for international distribution. The Disney+ deal offers a good example. Reportedly, TBS entered into the agreement, spurred by the realization that its own Paravi platform had failed to take off domestically and moreover that, having entered the streaming market late, it had little hope of matching the established global infrastructure of US-based platforms.[67]

TBS enters the agreement, however, largely at the mercy of Disney and its multiple agreements and commitments around the globe. While TBS has promoted it as a deal with Disney+ to distribute its dramas worldwide, the dramas are in fact only available in certain countries, due to the limited

---

65. TVer, "Minpō terebi no chijōha riarutaimu haishin ga TVer ni seizoroi! 2022 nen 4 gatsu 11 nichi (getsu) yoru kara sutāto!" (Realtime streaming of commercial television networks' broadcasts comes in full force to TVer! Starting on the evening of April 11, 2022!), press release, April 8, 2022, https://tver.jp/_s/info/TVer_release_realtime_20220408.pdf.

66. "CM shūnyū—shichōritsu genshō, TVer nado haishin kōkoku shūnyū zōka: Minpō kīkyoku dai 1 shihanki kessan" (Television advertising revenue and viewer ratings decrease, advertising revenue increases on TVer and other streaming: First quarter financial reports of key television networks), *Mainabi nyūsu* (Mynavi news), August 5, 2022, https://news.mynavi.jp/article/20220805-2417354/.

67. Inoue, "TBS no 'Netflix dokusen haishin.'"

availability of the STAR hub in whose catalog the dramas are included. Disney launched the hub in early 2021 in locales that included certain European countries, Canada, Singapore, and Australia, and expanded it to others, such as Japan, Hong Kong, and South Korea, in late 2021. Intended as an international alternative to Hulu, in whose US operation Disney still holds a majority stake, the hub traded on the familiarity in many countries outside the United States of the STAR brand, which Disney obtained as part of its 2019 acquisition of 21st Century Fox. Disney has held off including the hub on its US Disney+ service, apparently due to its continuing commitments to Hulu in the United States and on which Disney supplies much of the same content from its US subsidiaries, such as FX and ABC, that it does on the STAR hub outside of the United States.[68] In consequence, the TBS dramas included in the Disney+ agreement are currently not available in the United States through the streamer, and it is unclear when if ever they will be.

The Lasting Impact of Network Histories and the Future of Japanese Dramas

The Japanese government's aim for Japan to become an IP-oriented nation has, in the case of dramas at least, had mixed results, with Japan resembling more an IP fiefdom within a larger network of non-Japanese platforms. As already noted, these initiatives have done little to stanch the availability of pirated content outside of Japan, and, arguably, they have left Japanese media companies more beholden to, rather than independent from, the interests and demands of non-Japanese companies. The fate of Japanese dramas in the age of streaming underscores this, as the hodgepodge map of

---

68. Manori Ravindron, "Disney Lifts Lid on Star: Exclusivity and Parental Control Keys to New Tile," *Variety*, February 17, 2021, https://variety.com/2021/streaming/global/disney-plus-star-launch-1234909210/.

access offered by agreements like the TBS–Disney one highlights. Namely, even as international, generally US-based streamers offer a possible route for international exposure through channels that nominally protect Japanese IP, they do so in a manner that leverages IP to their own advantage and profit, rather than those of the content's rights holders.

One discerns in this unbalance the contours of what Dal Yong Jin has identified as "platform imperialism." As defined by Jin, platform imperialism instantiates "an asymmetrical relationship of interdependence" between the West—the US in particular—and other nations, sustained by the "technological and symbolic domination of US-based platforms."[69] Although platform imperialism as such extends the earlier logics of cultural imperialism, according to Jin, it also represents somewhat of a departure, given the extent to which "IP, entrepreneurship, and values [are] embedded in platforms."[70] As Jin has noted, this definition of platform imperialism nonetheless applies well to Netflix, as it underscores the manner in which the SVOD service leverages its data-based approach to assert itself within local markets, where it subsequently influences cultural production.[71] To this, one might add how the company has, in many ways, established the IP-management model that serves as a blueprint for global media distribution via streaming and how this further extends a US-centric approach to IP. In this regard, it is notable that US-backed trade agreements often incorporate and impose IP and other legal frameworks that mirror US ones and are thus favorable to the US technology companies familiar with them. Japan is no stranger to these issues: its government passed revisions to existing copyright laws in 2016 so as to conform to the anticipated requirements of the Transpacific Partnership—requirements that the United States, which ultimately pulled

---

69. Jin, *Digital Platforms*, 12.
70. Jin, 12.
71. Dal Yong Jin, "Netflix's Corporate Sphere in Asia in the Digital Platform Era," in *The Routledge Handbook of Digital Media and Globalization*, ed. D. Y. Jin (New York: Routledge, 2021), 167–75.

out of the agreement, had pushed for so as to align the pact's IP regime with its own.[72]

The underwhelming fate of Japanese dramas in the era of streaming highlights, however, how the domestic side of such asymmetrical relationships complicates an understanding of them as necessarily one-directional. Instead, it reveals the local contexts that shape platform imperialism's specific manifestations. As the counterexample of anime underscores, it is not sufficient to conclude that the difficulties faced by Japan's media companies in areas like scripted dramas is simply the result of an IP regime imposed from the outside. Rather, it is more accurate to say that such imbalances and difficulties arise from the intersection of externally imposed regimes and local structural particularities, such as extant industry practices and expectations. In the case of Japanese dramas, this intersection has not only played a role in creating a gap between Japan's domestic media industry and the transnational media systems with which it interacts; it has informed the split between the dramas consumed domestically and those most typically exported abroad for distribution via streaming.

---

72. Terakura Kenichi, "TPP to chosakukenhō kaisei: Kenri hogo to riyō no tekisei na kinkō o mezashite" (TPP and copyright law revisions: Aiming for a balance between rights protection and appropriate use), *Chōsa to jōhō* (Survey and information) 922 (2016): 1–12; "Trans-Pacific Partnership Agreement," Electronic Frontier Foundation, accessed March 15, 2023, https://www.eff.org/issues/tpp; David McCabe and Ana Swanson, "U.S. Using Trade Deals to Shield Tech Giants from Foreign Regulators," *New York Times*, October 7, 2019, https://www.nytimes.com/2019/10/07/business/tech-shield-trade-deals.html.

Transmedia Adaptation, Sonic Affect, and Multisensory Participation in Contemporary Chinese *Danmei* Radio Drama

YUCONG HAO

Abstract

Danmei, a genre also known as boys' love, first developed in China in the 1990s under the influence of Japanese subculture in the 1990s, but it has diverged from its Japanese antecedents in the last decade. Recent Chinese cultural products of danmei, no longer confined to a subcultural group, have attracted mainstream attention and been widely adapted into a variety of popular media forms. In this paper, with the case study of radio drama *Grandmaster of Demonic Cultivation* (2016), I examine how queerness is explored and experienced through the affective expressivity inherent in voice as well as the interactive interface of *danmaku*, with which listeners can experience and articulate queer fantasy in a relatively unobstructed way. As the genre of danmei has been subject to persistent state censorship, this paper further explores strategies of containment and tactics of negotiation deployed by both content producers and cultural consumers.

Keywords: danmei, radio drama, vocal timbre, transmedia adaptation, contemporary China

Danmei 耽美, a translingual loanword from the Japanese term *tanbi*, refers to a type of fictional representation that depicts romantic relationships between attractive male characters and is commonly known as boys' love (BL).

https://doi.org/10.3998/gs.3771

In China, danmei culture first developed in the 1990s with the import of Japanese subculture,[1] but it has diverged from its Japanese antecedents in significant ways in the last decade. Recent Chinese cultural products of danmei, no longer confined to a hobby consumed by subcultural groups, have attracted mainstream attention and been widely mediated and adapted across a variety of media forms.

In this paper, I trace the transmedia adaptation of danmei literature in popular media with the focus on the form of radio drama. While the phenomenon of adapting danmei transmedially testifies to the generic elasticity of this literary genre, each media form utilizes their medium specificities and negotiates with state regulation and platform interfaces to manufacture a distinct aesthetic and affect of queer intimacy. With the case study of radio drama *Grandmaster of Demonic Cultivation* released on audio platform Miss Evan, this paper illustrates how queer fantasy is explored and experienced through the affective expressivity inherent in voice as well as the interactive interface of *danmaku*. As the genre of danmei has been subject to persistent state censorship, this paper further explores strategies of containment and tactics of negotiation deployed by both content producers and cultural consumers. While sexual or romantic elements are modified or eliminated, transforming depictions of queer desire into "bromantic" homosociality,[2] avid danmei fans nevertheless deploy skills of close listening and collaborative storytelling to recover such narrative compromise.

---

1. See Jin Feng, "'Addicted to Beauty': Consuming and Producing Web-based Chinese 'Danmei' Fiction at Jinjiang," *Modern Chinese Literature and Culture* 21, no. 2 (2009): 1–7; and Ling Yang and Yanrui Xu, "Chinese Danmei Fandom and Cultural Globalization from Below," in *Queer Fan Cultures in Mainland China, Hong Kong, and Taiwan*, eds. Maud Lavin, Ling Yang, and Jing Jamie Zhao (Hong Kong: Hong Kong University Press), 3.

2. See Eve Ng and Xiaomeng Li, "A Queer 'Socialist Brotherhood': *The Guardian* Web Series, Boys' Love Fandom, and the Chinese State," *Feminist Media Studies* 20, no. 4 (2020) 479–95; Tingting Hu and Cathy Yue Wang, "Who Is the Counterpublic? Bromance-as-Masquerade in Chinese Online Drama—*S.C.I. Mystery*," *Television & New Media* 22, no. 6 (2021): 671–86; Angie Baecker and Yucong Hao, "Fan Labour and the Rise of Boys' Love TV Drama in China," *East Asia Forum Quarterly* 13, no. 2 (2021): 17–20.

From Subculture to the Mainstream

Borrowed from *tanbi*, a Japanese word meaning "addictions to beauty," danmei designates fictional representation of queer romance between male characters.[3] With the importation of Japanese BL comics in the 1990s, especially through media piracy and with the mediation of licensed translation in Taiwan, danmei culture flourished in the Chinese mainland, often with the cyberspace as its fan base.[4] Fan sites of Japanese BL and forums on danmei literature have been created since the late 1990s,[5] in which danmei readers expressed their enthusiasm for the genre by engaging in discussion, translation, and fan-fiction writing of danmei comics and literature. This initial stage of Chinese danmei was developed largely under the influence of Japanese boys' love, when Japanese cultural products, especially comics and novels, were favorably consumed and served as a model for the localization of BL in China.[6] This transnational flow of danmei attests to what Iwabuchi Koichi conceptualizes as the regional globalization in East Asia, in which Japanese popular culture circulates and inspires "familiar but different modes of Asian indigenized modernities in both cultural production and consumption."[7]

From the very beginning of danmei culture in China, the Internet serves as the primary cultural venue for Chinese fans, predominantly female, to build interest-based groups and exchange queer fantasies. The relatively light

---

3. Feng, "'Addicted to Beauty,'" 5.
4. Yang and Xu, "Chinese Danmei Fandom and Cultural Globalization from Below," 4–5; Chunyu Zhang, "Loving Boys Twice as Much: Chinese Women's Paradoxical Fandom of 'Boys' Love' Fiction," *Women's Studies in Communication* 39, no. 3 (2016): 249–50.
5. Popular *danmei* sites in this period include Sunsun Academy (founded in 1998 and closed in 2015), Lucifer Club (1999–), and Jinjiang Literature City (2003–), which continues to be the largest online community for danmei literature.
6. Xu and Yang, "Forbidden Love: Incest, Generational Conflict, and the Erotics of Power in Chinese BL Fiction," *Journal of Graphic Novels and Comics* 4, no. 1 (2013): 31–32.
7. Koichi Iwabuchi, *Recentering Globalization: Popular Culture and Japanese Transnationalism* (Durham, NC: Duke University Press, 2002), 18.

regulation of the Internet compared to print media, along with the freedom that anonymity offers, enables danmei fans to explore queer sociality and intimacy in ways that are otherwise impossible in a deeply heteronormative society. Some scholars hence identify these female fans as counterpublics, whose engagement with danmei culture "challenge mainstream heteronormativity in a liberating manner."[8] A contending view, however, suggests that despite their presumed interest in queerness, danmei fans are predominantly interested in the idealized, romanticist representation of male-male relationships while concerning themselves little with the realistic representation or real condition of homosexuals in China.[9]

In the twenty-first century, especially since the 2010s, Chinese danmei culture has made remarkable developments in formal and thematic diversity and metamorphosized from a subcultural, underground interest into a lucrative cultural enterprise that receives mainstream visibility. No longer a cultural derivative of or local response to Japan's regional influence, it actively engages with and contributes to the changing mediascape of the Chinese contemporary culture industry. In Ling Yang and Yanrui Xu's research, they observe that the field of Chinese danmei now consists of three prominent groups: the original danmei circle of Chinese-language fiction, the Japanese circle that consumes and re-creates Japanese BL, and the Euro-American circle that focuses on Western media and slash culture. While each of the circles has their own trajectory of development and consumes different media products, the boundaries between these circles are relatively fluid and porous, making cultural cross-fertilization a common phenomenon.[10]

---

8. Feichi Chiang, "Counterpublic but Obedient: A Case of Taiwan's BL Fandom," *Inter-Asia Cultural Studies* 17, no. 2 (2016): 224. See also Hu and Wang, "Who Is the Counterpublic?," 671–86; Yang and Xu, "*Danmei, Xianqing*, and the Making of a Queer Online Public Sphere in China," *Communication and the Public* 1, no. 2 (2016): 251–56; Jiang Chang and Hao Tian, "Girl Power in Boy Love: Yaoi, Online Female Counterculture, and Digital Feminism in China," *Feminist Media Studies* 21, no. 4 (2021): 604–20.

9. Chiang, "Counterpublic but Obedient," 228–33.

10. Yang and Xu, "Chinese Danmei Fandom and Cultural Globalization from Below," 8.

The popular consumption of domestic and Euro-American media products as well as extensive contacts with an international BL community (in particular, through the fan-fiction site Archives of Our Own) significantly diversifies the field of danmei culture in contemporary China, in which Japanese BL is no longer the sole source of influence. Moreover, among these three circles, the original danmei circle develops most rapidly and has aggregated the largest fan base.[11] These original Chinese-language danmei titles are often serialized on major danmei literature sites, especially Jinjiang Literature City, on a subscription basis.

Original danmei novels have taken on a variety of themes, such as school romance, science fiction, martial arts, historical novel, and immortality cultivation, among others. The mechanism of subscription provides economic incentives for both the service provider and the author to produce original works more and faster, and platforms usually embed a commenting system that allows readers to interact with the author and provide immediate responses, and even inspiration, regarding the plot, characters, and style.[12] The system has thus enabled a form of literature-production that is often fan driven, in which readers are encouraged to play an active role in the creation and development of a story. It also generates data that quantifies reader engagement and helps the platform and authors to track the popularity of different genres and works, identify emerging trends, and make predictions about what subject may resonate with readers. Although some novels will eventually get published and circulated in print form, especially through book publishers in Taiwan,[13] these danmei sites still serve as the primary

---

11. Xi Tian, "More than Conformity or Resistance: Chinese 'Boys' Love' Fandom in the Age of Internet Censorship," *Journal of the European Association for Chinese Studies* 1 (2020): 192.

12. Discussion about the author-reader relationship in Chinese Internet literature platforms can be found in Yuyan Feng and Ioana Literat, "Redefining Relations Between Creators and Audiences in the Digital Age: The Social Production and Consumption of Chinese Internet Literature," *International Journal of Communication* 11 (2017): 2589–2600.

13. Cathy Yue Wang, "Officially Sanctioned Adaptation and Affective Fan Resistance: The Transmedia Convergence of the Online Drama *Guardian* in China," *Series—International Journal of TV Serial Narratives* 5, no. 2 (2019): 47.

venue for original literary works to circulate and reach an ever-increasing readership on the Internet.

As the cultural impact of original danmei literature continues to grow, some novels have garnered the attention of major entertainment companies and video-streaming platforms. In light of the substantial commercial potential of the danmei genre, production companies have shown interest in purchasing the rights to adapt danmei novels into other popular forms of media. Recognizing danmei as a marketable cultural asset, they often refer to it as *intellectual property* (IP), a term that is used in the Chinese context to highlight the legal rights and economic prospect of a given cultural work, especially Internet literature. Production companies rate a given IP by metrics of S, A, B, and C, with a S rating indicating the most marketable. In the IP economy, if a danmei novel is categorized as S, then the companies will invest heavily to transform the novel into other popular media, ranging from web drama and animation to mobile games. As web drama is a particularly popular one among all these media, production companies often recruit A-list actors to star in the web-drama adaptation.

While an IP is not the same as a media franchise—for essentially all transmedia adaptations of an IP still are based on and follow the original story line—the economic drive and commercial mechanism nevertheless operate in a similar logic as branding and franchising,[14] in which production companies adapt an IP across many media forms with the goal to enhance its visibility and profitability. In the transmedia migration of an IP, ideally, each of these mediums, by retelling the same story through "recreations, remakes, remediations, revisions, parodies, reinventions, reinterpretations, expansions, and extensions,"[15] would contribute to the unfolding of the story in their own ways and generate discrete attractions of entertainment. However, in the case of adapting danmei in China, the question of media

---

14. Henry Jenkins, *Convergence Culture: Where Old and New Media Collide* (New York: New York University Press), 19.
15. Linda Hutcheon, *A Theory of Adaptation* (London: Routledge, 2013), 181.

regulation inevitably arises, especially considering the liminal space that queer imagination occupies and the uneven degree of regulation that each media platform is subject to. Unlike danmei literature that does not require a license to serialize on the Internet, when the genre is adapted and disseminated into visual, audio, or haptic media, permissions from state authorities are usually required. Thus, explicit representation of queer sexuality has to be modified, revised, or removed in order to subordinate the story to the scrutiny of regulators. In a way, instead of each medium making its unique contribution to the unfolding of the original fiction, each of these popular media adaptations exposes processes of compromise and traces of ambivalence in an effort to contain explicit queer content.

In 2016, a web drama series called *Addicted*, adapted from Chai Jidan's novel *Are You Addicted?*, aired on iQiyi and Tencent TV, two of the largest video-streaming platforms in China. When the sexual intimacy between the male leads became too obvious, the web drama was quickly taken down by the National Radio and Television Administration (NRTA). Despite the ban, *Addicted* continued to circulate underground through piracy, and its extensive popularity and the rapid rise to fame of the actors fueled the confidence of producers and assured them that there was an immense market in adapting danmei. The challenge that the production companies and streaming platforms had to grapple with, then, was how to negotiate between the popularity of the homosexual subject and regulation by the NRTA, which strictly censored cultural products depicting "abnormal sexual relations or sexual behaviour."[16]

Addicted is generally perceived as the prototype of web dramas adapted from danmei literature (*dangai ju*), and thus it is often referred to as *dangai* 1.0. Two years after, an adaptation of Chinese danmei author Priest's *Guardian* was released on Youku, another major video-streaming platform. Unlike *Addicted*'s explicit depiction of queer intimacy that makes the web

---

16. Baecker and Hao, "Fan Labour and the Rise of Boys' Love TV Drama in China," 18.

drama deemed promiscuous by the authority, the treatment of the same-sex relationship between the male leads in *Guardian* is much subtler. Portrayed as close colleagues and soul mates, the two protagonists undertake together the mission of safeguarding peace and order. Their queer relationship that is elaborately depicted in the original novel is now camouflaged as socialist homosocial brotherhood, though a more seasoned audience could hardly miss the romantic ties between the protagonists. *Guardian*, to a certain extent, reconciles the conflict between the popular appeal of homosexual romance and NRTA's restriction on queer content. Yet the drama was still removed from the streaming platform one week after the series concluded. It was only after a few scenes that contained "supernatural or harmful content" were deleted three months later that *Guardian* was made available again.

In 2019, web drama *The Untamed* was aired on Tencent TV. An immediate hit after its release, it broke new ground in danmei adaptations. The web drama was adapted from a novel by Mo Xiang Tong Xiu, *Grandmaster of Demonic Cultivation* (2016), which was serialized on Jinjiang Literature City from 2015 to 2016. A danmei fiction of immortality cultivation, the story revolves around two young lords, Wei Wuxian and Lan Zhan, who investigate dark happenings in the world of Daoist cultivation, and the central story line is concentrated on the protagonists' development of affection for each other. Just like *Guardian*, the queer romance between the protagonists is only alluded to in the web drama, and yet it still constitutes the central appeal of the show and captivates a large number of viewers. It is the most viewed web drama in 2019 with over four billion views, and by 2022, it has a total count of over ten billion views.[17] The web drama also helped the streaming platform develop paid services such as advanced on-demand screening, and it is estimated that the advanced viewing of the final episodes created a revenue of over 150 million yuan ($21,405,978 USD).[18]

17. Wang Yingmin, "Viewership of Chinese TV Dramas in 2019," Lianhe zaobao, January 13, 2020, https://www.zaobao.com/zentertainment/movies-and-tv/story20200113-1020684.
18. Fan Dongcheng, "Dangai ju at the Crossroads," The Paper, March 29, 2021, https://www.thepaper.cn/newsDetail_forward_11932215.

Prior to *The Untamed*, most dangai ju targeted a small audience of danmei fans. What makes *The Untamed* truly exceptional is how it has been simultaneously recognized by the mainstream media. On June 28, 2019, the overseas edition of *People's Daily*, a major state media outlet for overseas publicity, published a review on this web drama. Entitled "*The Untamed*: Writing the Beauty of Chinese Style," the essay lauds the web drama's extensive incorporation of Chinese traditions, which demonstrates "not only the shape of traditional culture, but the soul of the Chinese nation (*Zhonghua minzu*)."[19] In this way, the reviewer applauds *The Untamed* for its efficacy in "expressing cultural confidence" and "cultivating positive values." The review's curious, and perhaps intentional, neglect of the plot of queer romance that constitutes the mainstay of the story encapsulates the core paradox of danmei in contemporary Chinese mediascape: while the state recognizes the effectiveness of the genre to promote -mainstream values and enhance soft-power publicity, the nonheteronormative origin of the genre has to be concealed to contain any transgressive potentials.

Transmedia Adaptation: From Text to Voice

Aside from the immensely popular web drama adaptation *The Untamed*, *Grandmaster of Demonic Cultivation* also appears in animation, radio drama, and mobile games. The transmedia circulation of the story proves to be a successful marketing strategy that promotes the prominence of the IP, and it also attests to the formal elasticity of the danmei genre to be molded across different media forms and generate discrete experiential pleasure. In the following section, I examine the radio drama adaptation of *Grandmaster of Demonic Cultivation* released on the audio platform Miss Evan. This is considered by many fans the most authentic adaptation of the original novel. Compared to its web

---

19. Hu Xin, "*The Untamed*: Writing the Beauty of Chinese Style," *People's Daily* (Overseas Edition), June 28, 2019, http://culture.people.com.cn/n1/2019/0628/c1013-31200814.html.

drama counterpart, I argue that the affective expressivity inherent in voice and the interactive interface of the radio drama platform enables listeners to engage with and articulate queer fantasies in a relatively unobstructed manner.

Radio drama, also known as audio drama, was a popular cultural form widely consumed in China in as early as the 1970s. In a time when TV sets were still largely unavailable to ordinary households, radio structured people's daily activities and functioned as the primary medium of receiving information. As Nicole Huang's study of auditory culture in 1970s China shows, edited films were widely broadcasted in radio programs, which contributed to the sensory pleasure of the everyday amid a totalizing revolutionary soundscape.[20] Many films were repurposed for radio broadcasting, in which original film recordings were heavily edited to fit into the format of radio storytelling, and a third-person narrative was often inserted to provide contextual information for radio listeners.

In the subsequent two decades, radio drama continued to flourish in the Chinese mediascape: the popularity of edited films persisted, and there were also radio broadcasting of literary classics and serialized novels.[21] While radio drama became marginalized by newer, audiovisual mediums of TV, film, and the Internet since the 1990s, the creation of mobile audio platforms, such as Himalaya FM (*Ximalaya*) and Miss Evan (*mao er*), by making listening a much more portable experience, rekindled users' passion for audio cultural products.[22] Moreover, the emergence of professional voice-acting studios as well as amateur enthusiasts also provides essential talents for the popularity of radio drama.[23]

---

20. Nicole Huang, "Listening to Films: Politics of the Auditory in 1970s China," *Journal of Chinese Cinemas* 7, no. 3 (2013): 187–206.

21. Ke Xu, "Cat and the Plumber: An American Radio Drama in China," *Comparative Literature Studies* 57, no. 3 (2020): 421.

22. Joe Yizhou Xu and Jeremy Wade Morris, "App Radio: The Reconfiguration of Audible Publics in China Through Ximalaya.FM," in *Engaging Social Media in China*, eds. Guobin Yang and Wei Wang (Lansing: Michigan State University Press, 2021), 6–7.

23. Since 2013, a number of dubbing studios have been founded to specialize in radio drama as well as TV drama and animation dubbing. Leading studios include TrioPen, Voice Bear, and Voicegem. Many of them have received professional training in the dubbing department of film academies, but there are also a number of self-taught voice actors.

In Linda Hutcheon's classical study of adaptation, when discussing adaptations presented in the form of radio drama, she considers that this audio medium "brings the importance of the aural to the fore."[24] Therefore, radio drama, as an aural form of performance, requires that "each character/ voice must be aurally distinguishable, there cannot be too many of them. . . . Music and sound effects are added to the verbal text to assist the imagination of the listener."[25] In other words, in radio drama, as the vocal performance of a literary text, the differentiation of different voices—those of the characters and the narrator—as well as the addition of sound effects are of pivotal importance for listeners to follow the story line and understand the characters. In my discussion of the radio drama adaptation of *Grandmaster of Demonic Cultivation*, I seek to bring to light yet another aspect of vocal performance: vocal timbre, which constitutes and cements the affective appeal of danmei fantasy.

Radio drama *Grandmaster of Demonic Cultivation* was produced by TrioPen Studio, a dubbing studio founded in 2014 that specialized in voice acting. Unlike the original fiction that employs a third-person narrator, TrioPen's adaptation consists solely of dialogues between the characters, who address each other in the first person. This change, by making voice a central identifier of characters, invites listeners to pay attention to not only the semantic expression of language but the timbral quality of each individual voice. Although the function of differentiation, as Hutcheon explicates, continues to matter in the process of storytelling, it is equally important to attend to what kind of listening experience, auditory and affective, that vocal timbre can afford for the listeners. Here, I deploy the concept of vocal timbre to understand the sensory and affective experience when listening to the texture of voice. Vocal timbre, or *Klangfarbe* ("tone-color") in German, refers to the vocal quality of the voice that extends beyond semantic description or musical notation. As Andrew Jones theorizes, timbre encapsulates

---

24. Hutcheon, *A Theory of Adaptation*, 41.
25. Hutcheon, 42.

and reflects the acoustic trace of the resonating body that produced it. As an embodied and sensory artifact, timbre possesses irreducible historical specificities that are "indexically linked to the people, the instruments, or even the locales and moments that produced them."[26] By acknowledging and listening to the extrasemantic quality of timbre, we can grasp the intricate ways in which the human voice communicates and conveys affective experiences.

In *Grandmaster of Demonic Cultivation*, vocal timbre, rather than make each character a disembodied voice emanating from the abyss of the ethernet, indexes the actually voicing body of the speaking subject. The corporeality of timbre hence allows listeners to imagine the characters as concrete beings with whom they can embody their danmei fantasy. One listener relays their experience of listening to the vocal timbre in the comment section of the radio drama, writing, "When I listen to the radio drama, I don't feel at all that this is voice acting. It is entirely the real person of Wei Wuxian speaking. [The performance of voice], full of emotions and incredibly versatile, engenders a strong sense of immersion."[27] The comment, by focusing exclusively on the timbral quality of voice, captures the affective appeal of radio drama listening: both the sonic pleasure of voice and the auditory experience of total immersion. Receiving a considerable number of upvotes, the message is representative of the shared auditory experience that many listeners receive. It is, therefore, hardly surprising that many fans of danmei radio drama feel strong attachment to the vocal timbre of vocal performers, which they associate with certain characters, and production companies often employ the same group of voice actors to perform characters in transmedia adaptations to strengthen such vocal identification and sonorous pleasure.[28]

---

26. Andrew F. Jones, *Circuit Listening: Chinese Popular Music in the Global 1960s* (Minneapolis: University of Minnesota Press, 2020), 173.

27. ddl shi yuandongli, "Kudos to Mr. Lu," posted in the comment section of *Grandmaster of Demonic Cultivation*, season 1, episode 10, on Miss Evan, August 6, 2018.

28. For instance, the web drama version of *Grandmaster of Demonic Cultivation* and the animation adaptation employed the same set of voice actors to dub main characters.

In an interview on the production of radio drama *Grandmaster of Demonic Cultivation*, Cai Mao, chief operating officer of Miss Evan, recalled the process in which they collaborated with dubbing studios to find suitable voice actors. The vocal timbre of voice actors, according to her, served as the primary criterion in the selection process.[29] In fact, when danmei radio dramas recruit voice actors, rather than prioritize acting experience or acting skills, they place much emphasis on vocal timbre. In job advertisements posted on Mohello, a Chinese-language online community of voice acting, danmei radio drama producers often have specific descriptions of vocal timbre that they intend to look for. Popular labels, for instance, include "fragile *uke* voice" and "alpha *seme* voice."[30] Here, vocal timbre indexes both sound's expressive capacity of embodying the personality of a given character and its effectiveness of conveying sexual intimacy in a queer relationship.

As radio drama makes voice—in particular, dialogues between characters—the primary medium of expression, the polysemic, elusive, and extrasemantic nature of voice makes it an effective way in evading censorship. While it is convenient to censor images or text by locating depictions of obscene content, it is, by contrast, much more challenging to identify the transgressiveness of voice, especially when voice is used for its nonsemantic quality. The radio drama *Grandmaster of Demonic Cultivation*, when it comes to scenes of queer intimacy, often deploys the sound of whispers, breathing, and vibrato quaver to allude to the romantic encounter between the male leads. The absence of direct verbal expression serves the dual purpose of setting up barriers for regulators to track transgressive content and allowing listeners to freely imagine queer intimacy.

---

29. Shao Maomao, "Radio Drama Grandmaster of Demonic Cultivation Having Been Played for over 20 Million Times," *Jiemian News*, July 20, 2018, https://www.jiemian.com/article/2324030.html.

30. *Uke*, literally meaning "to attack," refers to the top in a homosexual relationship, whereas *seme*, "to receive," indicates the bottom position. Advertisements looking for voice actors can be found under the recruitment section on Mohello, https://www.mohello.com/zhaomujijiehao

In the original novel of *Grandmaster of Demonic Cultivation*, there is a classical episode in which Lan Zhan, in his complete drunkenness, reveals his affection for Wei Wuxian. In the web drama adaptation, Lan's confession, along with the intimate interaction between the male protagonists, is completely taken out, and instead, the drama opts to show the bromantic devotion between the two. The budding affection is only vaguely suggested through close shots of their eye contacts. The radio drama, by contrast, preserves the queer relationship in a much more elaborate manner. In the same scene, the voice actor playing the character of Lan Zhan deployed rhythmic variations in his speech, such as delay, pause, and the breathiness of voice, which manifested both the physical state of drunkenness and the romantic ambience of affectionate intoxication between the two. Through these nonverbal vocal performances, the representation of the queer interaction becomes legible and profoundly enjoyable to listeners.

Moreover, when listening to the radio drama, danmei fans often use a technique that can be approximated as "close listening." As Charles Bernstein has conceptualized it, close listening refers to a practice in which listeners consciously attend to the vocal performance of a text and attune themselves to the dynamic interplay between sound and semantics.[31] Through the practice of close listening, listeners become aware of the ways in which voice acting, vocal timbre, intonation, rhythm, and other sonic qualities shape and enrich the expressivity of the text. Many listeners of the radio drama *Grandmaster of Demonic Cultivation* exhibit a heightened degree of intermedial attentiveness to the textual and sonic representation of the story. In one testimony, for instance, a listener recounts the revelatory experience that they had when listening to the radio drama. Much more than simply reading out loud the original text, voice actors, with their vocal performance of the prolongation of syllables, delays between words, and change of intonation, effectively convey the complexity of the

---

31. Charles Bernstein, *Close Listening: Poetry and the Performed Word* (New York: Oxford University Press, 1998), 4–7.

psychology of the characters, which, as the listener comments, would otherwise go unnoticed when they read the novel.[32] Such reflection illustrates the intertextual and intermedial process of listening that avid fans perform, in which they demonstrate a keen awareness of the medium specificity of radio drama, and when listening, they skillfully integrate their sensory experiences of reading and listening to enhance the pleasure of consuming the danmei genre.

Danmaku as Collaborative Storytelling

Since 2018, radio drama *Grandmaster of Demonic Cultivation* was serialized exclusively on Miss Evan and became an immense hit, with each of the paid episodes garnering more than one million play counts. By 2022, it is the most-played radio drama on the platform. Miss Evan, founded in 2010, is one of the largest audio-streaming platforms in China that provides audio products ranging from popular music to autonomous sensory meridian response (ASMR). It is particularly known as the hub of Chinese-language radio dramas adapted from Internet literature and, in particular, danmei literature. Aside from *Grandmaster of Demonic Cultivation*, Miss Evan has purchased copyrights of a few other popular BL titles, and these radio dramas are among the most popular paid products on the platform.

Rather than broadcasted through traditional radio stations, contemporary radio dramas are primarily produced by and circulated on Internet audio platforms and yet they are still commonly referred to as radio dramas (*guangbo ju*) by both audio platforms and listeners. The retainment of the name *radio* despite shifting distribution methods, on the one hand, helps to position these digital audio products within the genealogy of radio drama

---

32. ying fei zhu ri, "On Wei Chao's Dubbing of Lan Zhan," posted in the comment section of *Grandmaster of Demonic Cultivation*, season 1, episode 5, on Miss Evan, September 16, 2019.

in the twentieth century and illuminates how audio storytelling persistently captivates listeners through the use of sound effects, music, voice acting, and vocal timbre. On the other hand, it highlights the presence of state media regulation and control that are reminiscent of the practice of traditional radio broadcasts, which continue to loom over the production and consumption of Internet-based radio dramas in the contemporary Chinese mediascape.[33]

In the study of media in contemporary China, it is almost impossible not to address the censorship of media contents by state authorities. Yet preoccupation with state regulation risks assuming a simplistic binary between state control and individual agency, oppression and resistance, which may lead one to overlook the murky terrain of media politics triangulated by regulators, platforms, and users. Researchers have recently advocated ideas such as self-regulation or delegated censorship to highlight the dynamic, multilevel, and multilayered process of censorship, one "with a hardened core but flexible periphery."[34] This renewed view maintains that, rather than blunt censorship that directly comes from the state, Internet platforms moderate media content with relative autonomy and based on their own interpretations that do not align squarely with the demand of the state. In the case of Miss Evan, such incongruity between the state and the platform is discernible at several levels, from the interface design of the mobile app to the embedding of an interactive commenting system of danmaku. While the state is apprehensive about the transgressive potential of queer culture, the platform, seeing the impressive revenue that this genre can generate, cautiously promotes danmei radio drama while ensuring compliance with regulations.

Unlike other popular streaming services that use algorithms to display personalized content to users, Miss Evan features a navigation menu that

---

33. Xu and Morris, "App Radio," 9.
34. Taiyi Sun and Quansheng Zhao, "Delegated Censorship: The Dynamic, Layered, and Multistage Information Control Regime in China," *Politics & Society* 50, no. 2 (2022): 194. See also Mengying Li, "Promote Diligently and Censor Politely: How Sina Weibo Intervenes in Online Activism in China," *Information, Communication & Society* (2021): 1–16.

allows listeners to browse radio dramas by subject, status, and update schedule. Yet far from providing the liberty for listeners to explore freely, the app deploys a popularity ranking system that promotes and solidifies the visibility of audio dramas that have already aggregated considerable play counts. In terms of subject, radio dramas are subdivided into five categories on the platform: romance, modern, ancient style, Republican era, and others, and unsurprisingly, there is not a separate group for danmei.

These categories may seem confusing, as they represent distinct criteria, with some denoting genre while others indicating time periods. A closer examination of radio dramas under each of the categories, however, reveals the underlying logic: whereas romance refers exclusively to heterosexual love stories, popular titles under the remaining four categories are predominantly danmei radio dramas, which are grouped by time periods from ancient settings to contemporary. Though the platform carefully conceals the subject of danmei that a keyword search for the genre would yield no result, the classification system makes it handy for avid listeners to distinguish heterosexual romance from queer content and locate danmei radio drama of their preference.

Whereas the platform Miss Evan straddles between the promotion of danmei radio dramas and the concealment of the danmei genre, it incorporates the function of danmaku ("bullet screen") within its audio-playback interface that allows listeners to present their discussion and interpretation of queer story lines and relationships in a more unobstructed manner. Despite the ambivalent stance of the platform, the introduction of the danmaku interactive system provides an essential digital infrastructure for listeners of danmei audio drama to express and celebrate their queer fantasy and forge communal belonging.

Danmaku is an interactive commenting system first offered by Japanese online video site Nico Nico, which enables real-time comments to fly across the video like bullets barraging the screen. Combining "visual content of moving images with paratextual information of peer interpretations and feedback," danmaku transforms the experience of video viewing into that of social communication.[35] In China, this commenting interface has been popularized by video platforms Bilibili and ACFun since the 2010s and was later adopted by major video-streaming sites and was even introduced to film exhibition in the movie theater. Existing studies of danmaku in the contemporary Chinese mediascape tend to understand the phenomenon as a form of social interaction in which the paratextual (danmaku) displaces the visual (video), and communication matters much more than content.[36]

The use of danmaku on video-streaming platforms is often characterized as playful or disruptive, in which danmaku either engenders visual spectacles that "barrage" video play or parodies the original content with techniques such as *soramimi* ("mishearing") to create comic effects.[37] However, the use of danmaku on Miss Evan is markedly different. Here, it serves more functional purposes of facilitating information transmission and encouraging feedback and discussion of listeners. Rather than being used for playful deconstruction or parody, danmaku on Miss Evan synthesizes the role of information and communication.

Danmaku posted on *Grandmaster of Demonic Cultivation* and, by extension, other danmei radio dramas on Miss Evan generally consist of the following types: (1) subtitling: to provide textual transcription of audio; (2) interpretation: to discuss plot, characters, and relationship; (3) supplement: to supply information that is edited out or deleted due to censorship of queer content; (4) expression: to reveal affection for characters or the queer relationship; and (5) community-building: to interact with other listeners

---

35. Jinying Li, "The Interface Affect of a Contact Zone: *Danmaku* on Video-Streaming Platforms," *Asiascape: Digital Asia* 4, no. 3 (2017): 235.

36. See Li, "The Interface Affect of a Contact Zone," 233–56; Xuenan Cao, "Bullet Screens (*Danmu*): Texting, Online Streaming, and the Spectacle of Social Inequality on Chinese Social Networks," *Theory, Culture & Society* 38, no. 3 (2021): 29–49.

37. Nakajima Seio, "The Sociability of Millennials in Cyberspace: A Comparative Analysis of Barrage Subtitling in Nico Nico Douga and Bilibili," in *China's Youth Cultures and Collective Spaces*, eds., Vanessa Frangville and Gwennaël Gaffric (New York: Routledge, 2019), 105–6; Zhen Troy Chen, "Slice of Life in a Live and Wired Masquerade: Playful Prosumption as Identity Work and Performance in an Identity College Bilibili," *Global Media and China* 5, no. 3 (2020): 332.

and create a sense of communal belonging. If danmaku on video sites encapsulates a competition between the textual and the pictorial for the attention of users' vision, to embed danmaku in the interface of audio playback, in contradistinction, signifies the process of mutual complement between listening and viewing. Danmaku on Miss Evan, particularly through functions of subtitling, interpretation, and supplement, enhances the listening experience and establishes a meaningful logic between the audiovisuals.

Soramimi, first developed in Nico Nico, also becomes popular in the danmaku culture in China. *Soramimi*, literally meaning "sky ears," describes the phenomenon of when the listener hears something that is not what is actually being said, as if it were coming from the sky. With soramimi, viewers mischievously mishear and send danmaku that replaces the original word with implausible homonyms to engender humor. Moreover, viewers often strive to distinguish their misheard homophones from each other to manifest individual wit and creativity. The practice, by contrast, is largely absent in danmei radio dramas. Instead, listeners aim to facilitate the listening experience with textual clarity, precision, and authenticity to the original. Signature lines, rather than being replaced or displaced by soramimi, are collaboratively posted and accurately repeated by listeners to barrage the screen, in which the density of the textual overlay functions as the barometer of their affective intensity. In a key scene in *Grandmaster of Demonic Cultivation*, when Lan Zhan declares that Wei Wuxian belongs to him, the screen is covered fully with the overlay in which listeners unanimously repost the same phrase uttered by the protagonist.

Even though audio storytelling poses a greater challenge to media regulators,[38] depictions of explicit queer sexuality are still removed on Miss Evan to prevent radio dramas from being taken down by state authorities. On those occasions, danmaku affords listeners the opportunity to discuss, interpret, and supplement the queer story line that has been moderated. In scenes

---

38. Fan Yang, "Feminist Podcasting: A New Discursive Intervention on Gender in Mainland China," *Feminist Media Studies* (2022): 11–12.

when protagonists interact intimately with each other, as the radio drama mainly deploys extraverbal expressions of quavers and gasping breaths to imply the strong affection developed between the male characters, many listeners opt to fill in the semantic gap by sending danmaku that cites the original text or presents close readings to make explicit the queer relationship. The mechanism of danmaku thus affords a collaborative form of storytelling, in which vocal performance of voice actors and visual-haptic response from listeners jointly engender the discrete pleasure of queerness.

In the realm of auditory culture, scholars often characterize listening as a highly private and intimately inward-looking experience.[39] When one listens to an audio product, the listener engages with the outside world by silently and solitarily mediating one's interior feelings and exterior narratives provided by the audio content.[40] Yet in the case of Miss Evan, the feature of danmaku complicates the practice of listening by simultaneously activating other senses—in particular, sensory experiences of the visual reading of scrolling comments and haptic feedback of typing and sending danmaku. The synthesis of senses contributes to the engendering of a more immersive and holistic listening experience, one that transforms the engagement with danmei radio drama from a private, contemplative auditory activity to a collective and interactive multisensory one. As is demonstrated in the analysis above, especially on occasions when the narrative of radio drama is partially compromised, danmaku proves to work effectively in that it provides listeners with a platform to reclaim queerness through the collective and multisensory labor of reconstructing the queer story line.

Writing on the use of danmaku in the process of community-building, existing scholarship reveals the illusory nature of such communal belonging

---

39. See, for instance, Jonathan Sterne, "Introduction," in *The Sound Studies Reader*, ed. Jonathan Sterne (New York: Routledge, 2012), 1–17; Susan J. Douglas, *Listening in: Radio and the American Imagination* (Minneapolis: University of Minnesota Press, 2004).
40. See Lukasz Swiatek, "The Podcast as an Intimate Bridging Medium," in *Podcasting: New Aural Cultures and Digital Media*, eds., Dario Llinares, Neil Fox, and Richard Berry (London: Palgrave Macmillan, 2018), 173–87.

that despite the impression of simultaneous viewing, there is no synchronic interaction among danmaku commentators.[41] In the case of danmaku on Miss Evan, the sense of simultaneity and its promise of real-time interaction matters less in the process of forging of a fan community. Instead, communal bondage and solidarity hinge upon a shared interpretive capacity of listeners of danmei radio drama to recognize the queer story line, reveal the implied, and supplement the lack. In other words, a type of danmei literacy performed in the process of listening and danmaku posting is indispensable to cultivating a sense of collective ownership of the cultural property and substantiating the sense of communal identity.

Listeners of danmei radio drama have demonstrated admirable sophistication in response to moderation and censorship. Showing a deep understanding of the cultural and sociopolitical ecology of contemporary Chinese media, they are well aware of the precarity of the genre, which prompts them to safeguard the genre from total elimination while avoiding direct confrontation with the logic of censorship. In their danmaku, listeners do not criticize the production team for complying with regulations, nor directly attack the NRTA for censoring queer content. Rather, taking censorship as a given condition and showing little willingness to challenge the status quo, they express appreciation and gratitude that the queer romance can still be partially presented.[42]

Conclusion

The present paper, with the case study of radio drama, offers preliminary research of the transmedia adaptation of danmei literature in contemporary China. The transmedial journey not only demonstrates the versatility

---

41. Li, "The Interface Affect of a Contact Zone," 247–51.
42. This phenomenon can be best illustrated by the finales of most danmei radio dramas, where listeners post predominantly *danmaku* to express their gratitude and respect to the production team who made the audio renditions possible despite challenges inherent to the genre.

of the genre across different media forms but brings to light their distinctive medium specificities and affective expressivity. With a focus on auditory and affective experiences of queerness that danmei radio drama affords, the study reveals the distinct sonic affect that the medium, platform, and interface of radio drama manufacture. In so doing, it moves beyond a plot-centered approach to Chinese danmei culture that often gravitates toward the narration and articulation of queer imagination, as well as a discursive framework that attends primarily to the media politics of queer culture.

The nonsemantic and extrasemantic elements, such as vocal timbre and the interactive interface of danmaku, exemplify how production companies, digital interfaces, and ordinary listeners negotiate between media regulations, economic revenues, and queer fantasy. Furthermore, the focus on sound media, an increasingly important medium in contemporary Chinese mediascape, illuminates the importance of understanding the medium affordance and media politics with which audio content is produced and consumed, and the vantage point of sound also reveals the potential of the sonic in shaping collective agency and communal belonging. While existing scholarship has focused heavily on the audio medium of podcast, the current study, by looking at danmei radio drama and the interfacing effects of danmaku, reveals participatory and multisensory experiences enabled by this audio-based form of communication.

Acknowledgments

I would like to thank Angie Baecker for her inspiration and encouragement, which are indispensable to the research and writing of this project. I would also like to thank the anonymous reviewers and the editors Lina Qu, Tze-lan Sang, and Ying Zhu for their generous support and feedback.

The Nostalgic Negotiation of Post-TV Legibility in *Mom, Don't Do That!*

EUNICE YING CI LIM

Abstract

The success and legibility of South Korean Netflix dramas like *Squid Game* (2021) have reinforced the position of Korean television dramas (K-dramas) at the forefront of both Asian and global entertainment. As one of the major consumers of Korean Wave products, Taiwan's entertainment industry necessarily situates itself regionally, transnationally, and globally in relation to and in response to K-dramas' dominance. The recently released Taiwanese Netflix television series *Mom, Don't Do That!* (媽,別鬧了!, 2022) is a key development in Global Taiwanese television drama that reflects on and responds to the unabating Korean Wave (Hallyu or K-Wave) and the regional and global popularity of K-dramas on Netflix and other post-television over-the-top (OTT) media-viewing platforms. By alluding to the genre conventions that characterize Taiwanese prime time Hokkien dialect (*Minnan*) dramas and Taiwanese Mandarin idol-dramas, *Mom Don't Do That!* (*MDDT*) strategically encourages a transnational nostalgia toward a golden era of Taiwanese dramas. This nostalgia, in turn, strategically allows *MDDT* to negotiate a new position for global Taiwanese television drama, challenging K-dramas' ubiquity and Netflix's curative influence on global media-consumption cultures. *MDDT* self-reflexively stages the domestic obsolescence of television, juxtaposes their screen-mediated food-consumption practices with the mediated foodways that have come to characterize the K-Wave, and highlights the unlikely and unhealthy nature of K-drama's romantic tropes and archetypes. These strategic moves variously establish *MDDT* as a sustained metacommentary on the K-drama-dominated, post-TV conditions that continue to shape media-viewing practices and cultures of today.

https://doi.org/10.3998/gs.3739

87

Keywords: Global Taiwanese television drama, Netflix, Hallyu, Korean Wave, Post-TV, Transnational nostalgia

The recent success and legibility of South Korean Netflix dramas like *Squid Game* (오징어 게임)[1] and *All of Us Are Dead* (지금 우리 학교는)[2] have reinforced the position of Korean television drama serials (K-dramas) at the forefront of East Asian and global entertainment. Having offered Korean television shows to its US members since 2012,[3] Netflix, the popular sub-scription video-on-demand (SVOD) over-the-top (OTT) streaming service, has enabled K-dramas to enjoy widespread circulation and recognition beyond East and Southeast Asia, where Korean entertainment has the advantage of cultural and geographical proximity. In recent years, Netflix has been expanding its collection of Taiwanese entertainment offerings, and this expansion includes Netflix-funded and distributed Taiwanese dramas like *Light the Night* (華燈初上)[4] and *The Victim's Game* (誰是被害者).[5] Taiwan is a major consumer of South Korean entertainment and popular culture or Korean Wave (K-Wave or Hallyu) content,[6] so much so that the

1. Dong-hyuk Hwang, *Squid Game*, performed by Jung-jae Lee, Hae-soo Park, Ha-joon, and HoYeon Jung, Netflix, 2021, https://www.netflix.com/title/81040344.
2. Jae-kyoo Lee and Nam-su Kim, *All of Us Are Dead*, performed by Chan-young Yoon, Ji-hu Park, Yi-hyun Cho, and Lomon, Netflix, 2022, https://www.netflix.com/watch/81237996/.
3. Hyejung Ju, "Korean TV Drama Viewership on Netflix: Transcultural Affection, Romance, and Identities," *Journal of International and Intercultural Communication* 13, no. 1 (2020): 33.
4. Yi-chi Lien, *Light the Night*, performed by Ruby Lin, Yo Yang, Cheryl Yang, and Rhydian Vaughan, Netflix, 2021–22, https://www.netflix.com/watch/81482855.
5. David Chuang and Kuan-chung Chen, *The Victim's Game*, performed by Joseph Chang, Wei-ning Hsu, Shih-hsien Wang, Netflix, 2020, https://www.netflix.com/watch/81230884.
6. Yu-Tien Huang and Jowon Park, "Taiwanese Korean Drama Viewing Experience through the Internet Prior to Official Import and Its Relationship with the Afterward Intention to View the Officially Scheduled Drama: The Case of 'The Legend of the Blue Sea' by SBS," *Journal of Media Economics & Culture* 15, no. 4 (2017): 162.

K-Wave has been credited with improving diplomatic relations between Taiwan and South Korea after the latter broke off diplomatic relations with Taiwan in 1992 to establish diplomatic relations with mainland China.[7] Considering the current dominance and popularity of K-dramas on Netflix, within Taiwan and around the world, Taiwan's entertainment industry necessarily situates itself regionally, transnationally, and globally in relation to and in response to K-dramas' dominance.

I argue that among these Netflix originals, the recently released television drama series *Mom, Don't Do That!* (媽別鬧了, henceforth abbreviated as *MDDT*)[8] is a key development in global Taiwanese television drama that strategically repositions Taiwanese dramas in relation to the popularity of K-dramas and the normalization of SVOD media-viewing cultures. Adapted from a Taiwanese bestselling novel,[9] the narrative of *MDDT* may initially appear to be a straightforward family drama and comedy reflecting on the everyday domestic and relationship troubles of an elderly widow named Mei-mei and her two unmarried adult daughters named Ru-rong and Ruo-min. Having lost Chen Guang-hui, Mei-mei's husband and the father of Ru-rong and Ruo-min, to a sudden heart attack five years prior, the three women of the Chen household continue to struggle to adapt to a life without their primary caregiver, companion, and mentor. Eager to remarry and desperate to marry off her two adult daughters, Mei-mei makes a bet with them to see who will get married first. Although this is *MDDT*'s deceptively simple and comedic premise, the production background of the Netflix original series already suggests that the seemingly lighthearted family drama might have more to offer than a casual viewer might expect. Backed

---

7. Sang-Yeon Sung, "Constructing a New Image. Hallyu in Taiwan," *European Journal of East Asian Studies* 9, no. 1 (2010): 25–27.

8. Wei-ling Chen and Chun-hong Lee, *Mom, Don't Do That!*, performed by Billie Wang, Alyssa Chia, Chia-Yen Ko, Johnny Kou, Kang-ren Wu, and Po-hung Lin, Netflix, 2022, https://www.netflix.com/watch/ 81477928.

9. Ming-min Chen, *My Mother's Foreign Wedding* [我媽的異國婚姻] (Taipei: Eurasian Press [圓神出版社], 2018).

by CJ E&M Hong Kong, the Southeast Asian headquarters of South Korea's CJ Entertainment & Media,[10] *MDDT*'s production background is suggestive of the inter-Asian and transnational dialogic ties and discursive connections being established between South Korean and Taiwanese entertainment industries. Enjoying the support of one of the largest and most prominent South Korean entertainment and mass-media companies, this easily negligible production detail complicates our understanding of global Taiwanese drama's uneasy, but also undeniable, connections with South Korea's entertainment industry. The respective globalizing pursuits and ambitions of Taiwan and South Korea's entertainment industries have, in other words, resulted in a series of strategic media convergences and divergences between their industries.

I argue that *MDDT* may be read as a dramatized metacommentary that reflects on and responds to the global dominance and popularity of K-dramas and post-TV media-consumption cultures. *MDDT* does acknowledge both the popularity and success of South Korean entertainment and the enabling role of Netflix in facilitating South Korea's and their own global production, circulation, and reception. Yet, it also offers commentary on the less desirable effects and implications of K-drama's increasing visibility and popularity, as it is facilitated by SVOD media-consumption cultures. By strategically offering subtle commentaries on K-dramas and post-TV online media consumption within the narrative of an ostensibly Taiwanese drama, *MDDT* deftly renegotiates a position for global Taiwanese television drama, seeking to renew its relationship with its audience through its consistent evocations of transnational nostalgia. In doing so, the drama discreetly challenges the ubiquity of both K-dramas and Netflix, drawing attention to the latter's curative influence and the prevailing effects such influence has on regional and global media-consumption cultures.

---

10. Karen Chu, "Netflix Comedy Series 'Mom, Don't Do That!' Wins Taipei Festival Award Ahead of Worldwide Debut," *Hollywood Reporter*, July 14, 2022, https://www.hollywoodreporter.com/tv/tv-news/netflix-comedy-series-mom-dont-do-that-wins-taipei-festival-award-ahead-of-worldwide-debut-1235180986/.

The Implied Authorial Audience: Recalling the Golden Era of Regional Taiwanese Dramas

The Taiwanese entertainment industry had been gaining its regional footing throughout the 1980s and 1990s, and its prime-time Hokkien-dialect (*Minnan*) television dramas and Mandarin idol-dramas were regionally popular in mainland China and Hong Kong, as well as in Southeast Asian countries like Malaysia and Singapore, throughout the 2000s. Minnan television dramas like *Dragon Legend: Fei Lung* (飛龍在天)[11] and Mandarin idol-dramas like *Meteor Garden* (流星花園)[12] and *Autumn's Concerto* (下一站，幸福)[13] had immense cultural influence and popularity in many parts of East Asia and Southeast Asia. The Taiwanese idol-drama adaptation of *Meteor Garden*, though originally inspired by a Japanese shōjo manga series,[14] was so well-received that it went on to inspire a Japanese live-action series adaptation,[15] a South Korean television series remake,[16] and a mainland Chinese television series remake.[17] This regional popularity of Taiwanese dramas in the 2000s—arguably a golden era of regional Taiwanese entertainment—coincided with and was subsequently disrupted by the parallel rising popularity of K-dramas, which continued to attract

---

11. Kai Fung, *Dragon Legend: Fei Lung*, performed by Nic Jiang, Alyssa Chia, Eric Huang, and Phoenix Chang, Formosa Television, 2000–1, Television Series.
12. Yueh-hsun Tsai, *Meteor Garden*, performed by Barbie Hsu, Jerry Yan, Vic Chou, Ken Chu, and Vanness Wu, Chinese Television System Inc., 2001, Television Series.
13. Hui-ling Chen, *Autumn's Concerto*, performed by Ady An, Vanness Wu, Ann Hsu, and Wu Kang-ren, Taiwan Television, 2009–10, Television Series.
14. Yōko Kamio, *Boys over Flowers* (Tokyo: Shueisha, 1992–2008).
15. Yasuharu Ishii, *Boys over Flowers*, performed by Mao Inoue, Jun Matsumoto, Shun Oguri, and Shota Matsuda, Tokyo Broadcasting System, 2005, Television Series.
16. Ki-sang Jeon, *Boys over Flowers*, performed by Hye-sun Ku, Min-ho Lee, Hyun-joong Kim, and Bum Kim, KBS2 and Netflix, 2009, Television Series, https://www.netflix.com/watch/70189961.
17. Helong Lin, *Meteor Garden*, performed by Shen Yue, Dylan Wang, Darren Chen, and Caesar Wu, Hunan Television and Netflix, 2018, Television Series, https://www.netflix.com/watch/81005506.

more viewership and accrue regional and global interest beyond the 2000s. This disruption has produced a palpable discontinuity or gap in the regional circulation and reception of Taiwanese dramas. Although Taiwanese dramas continue to be made, few dramas enjoy the enduring regional and international popularity and influence that Taiwanese television dramas in the 2000s did, and subtly calling attention to this discontinuity and reestablishing continuity today in the 2020s is what *MDDT* seeks to achieve.

References to the characteristic styles of both Minnan television dramas[18] and Mandarin idol-dramas in *MDDT* encourage a transnational nostalgia toward Taiwanese dramas of the early 2000s. By strategically establishing continuities and discontinuities between the 2000s era of Taiwanese dramas in a contemporary Taiwanese drama, *MDDT* seeks to reconnect with its audience from the 2000s era, reminding them of the reasons they used to enjoy Taiwanese dramas. Self-referentiality or self-reflexive devices have been identified as characteristic of Taiwanese idol-dramas. As has been previously theorized in relation to Taiwanese idol-drama, the "use of self-reflexive gestures that foreground artifice in a highly mimetic genre . . . serve[s] as a form of covert communication between implied author and authorial audience."[19] Michelle Wang distinguishes the "implied author" of Taiwanese idol-dramas from "the real or flesh-and-blood author," arguing that the implied author is "particularly useful in narrative genres like television drama serials, because these are not single-authored narratives but the synthesis of varied and multiple creative efforts on the parts of the

---

18. These *Minnan* dramas are known as 台語劇 or 鄉土劇 and were popular in China, Hong Kong, Malaysia, and Singapore, as well as other Southeast Asian countries, especially in the 2000s. While the domestic and overseas popularity of these *Minnan* dramas have undeniably been affected by the popularity of K-dramas, they remain popular among some groups of people in these places. In Singapore, for instance, *Minnan* dramas like *Dragon Legend: Fei Lung* (飛龍在天, 2000), *Taiwan Ah Cheng* (台灣阿誠, 2001), *The Spirits of Love* (愛, 2006), and *Night Market Life* (夜市人生, 2009) were and are still often the preferred prime-time television shows of many older Singaporean Chinese generations.

19. W. Michelle Wang, "[這又不是演戲] 'We're not playacting here,'" *JNT: Journal of Narrative Theory* 45, no. 1 (2015): 106.

director, screenwriter, actors and actresses." Building off Wang's identifying of an implied author in Taiwanese idol-dramas, I propose that in the case of *MDDT*, the implied author is the Taiwanese entertainment industry responding collectively to a post-TV entertainment industry dominated by K-Wave entertainment offerings. The implied authorial audience that this implied author self-reflexively reaches out to is one that is familiar with both Taiwanese dramas in the past as well as the popularity of K-dramas today, especially as they are both mediated through international entertainment circuits, online broadcast networks, and streaming services.

For instance, during a cab ride, Ru-rong—the stressed-out elder sister of the Chen family—starts to imagine her family members as stereotypical Minnan-speaking characters in a fictitious, melodramatic family drama titled *The Sound of Mother's Lament* (阿母的靠北聲).[20] Simulating an actual prime-time television drama playing on the small screen inside the cab and on a big outdoor screen that the cab passes, the stylized bright-red title is emblazoned on the top right-hand corner of the screen (see Figure 1), a characteristic visual format of Minnan television dramas.

This flitting and humorous sequence in the drama's representation of Ru-rong's financial and familial predicament serves as a nostalgic reference to this regionally popular genre, strategically reminding its implied authorial audience of how it was considered prime-time entertainment that undeniably contributed to the Taiwanese or Chinese wave in the 2000s. *MDDT* also references the regionally popular Minnan television drama *Dragon Legend: Fei Lung* by having Ru-rong encounter her former ex-boyfriend, a minor character played by Nic Jiang.[21] Since actress Alyssa Chia, who plays Ru-rong in *MDDT*, also plays Jiang's love interest in *Dragon Legend*, this interdrama connection is a delightful Easter egg for those who recall and recognize this on-screen romantic couple from more than twenty years

---

20. *Mom, Don't Do That!*, episode 3, performed by Alyssa Chia, Billie Wang, and Chia-Yen Kuo.
21. *Mom, Don't Do That!*, episode 1, performed by Alyssa Chia and Nic Jiang.

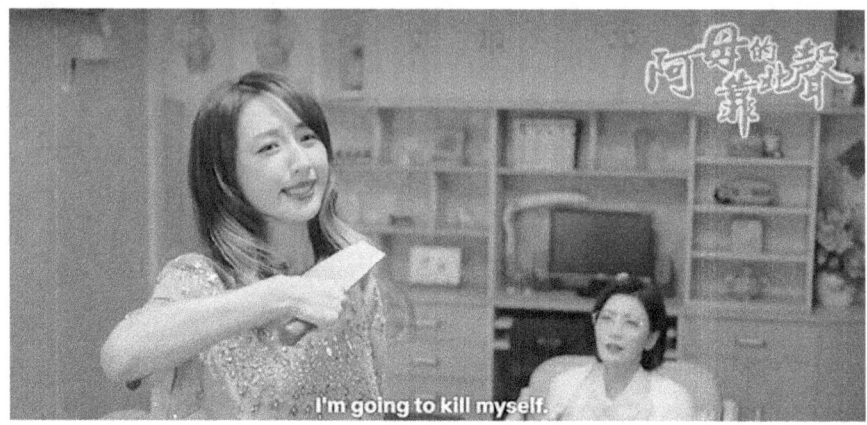

I'm going to kill myself.

Figure 1: In this fictitious Minnan television drama, Mei-mei and Ruo-min take turns dramatically holding a cleaver to their necks, threatening Ru-rong with killing themselves if she does not give them financial help. *Source*: Screenshot from the Netflix drama series *Mom, Don't Do That!* (媽,別鬧了!, 2022), episode 3

ago, at once attesting to *Dragon Legend*'s regional popularity and continued resonance. Jiang is not the only casting choice that reflects an intention to remind the implied audience of Taiwan's golden era of television. Boasting a stellar and beloved cast of veteran and popular Taiwanese actors and actresses like Billie Wang, Shao-hua Lung, Alyssa Chia, and Kang-ren Wu, *MDDT* adopts and adapts beloved and memorable elements and characters from its Minnan melodramatic family dramas and Mandarin romantic idol-dramas.[22] These adaptations strategically remind the audience of their relationship with Taiwanese entertainment, rewarding them for recalling these previously popular Taiwanese television genres and household names. The use of self-reflexive devices in *MDDT* thus reminds the audience of both Minnan family melodramas and Mandarin idol-dramas, representing

---

22. While it is more evident how *MDDT* is part *Minnan* family melodrama here, it will be evident later in my article how *MDDT* is also part romance-idol drama when I dive into Ru-rong and Senior's romance.

an amalgamation of the two popular Taiwanese television drama genres of the 2000s that is likely to evoke nostalgic reactions. At the same time, through these strategic borrowings, *MDDT* establishes itself as a new, hybrid genre of versatile global Taiwanese television drama that seeks to renew its relationship with the former audience of both Taiwanese Minnan television dramas and Mandarin idol-dramas.

The Spatial Relegation of Traditional Television to Nostalgic Clutter

It is within this hybrid genre of versatile global Taiwanese television drama that *MDDT* boldly stages television's obsolescence as a crucial subtext to the online dating platforms that Mei-mei uses, and this staging of television's obsolescence is also meant to evoke transnational nostalgia, which characterizes the domestic and emotional landscapes of the Chen household. When younger sister Ruo-min moves back home after catching her boyfriend Cha cheating on her, she is exasperated to discover that her mother and older sister have filled her room with numerous boxes of clutter in her absence, treating her room like their own storage facility[23] (see Figure 2).

This all-too-familiar and deceptively simple domestic scenario and premise at the start of *MDDT* motivates the many recurring scenes of the family packing and sorting through boxes throughout the course of the drama. In the process, the family discovers long-forgotten photographs, the personal diary of the late Guang-hui, an old tape recorder, and cassette tapes of popular Taiwanese music in the late '80s and early '90s.[24] These scenes of the family rifling through nostalgic clutter sets the stage for other

---

23. *Mom, Don't Do That!*, episode 1, performed by Chia-Yen Kuo, Po-Hung Lin, Billie Wang, and Alyssa Chia.
24. These cassette tapes specifically reference the Taiwanese boy band Little Tigers (小虎隊) and the girl group Yu Huan Pai Tui (憂歡派對).

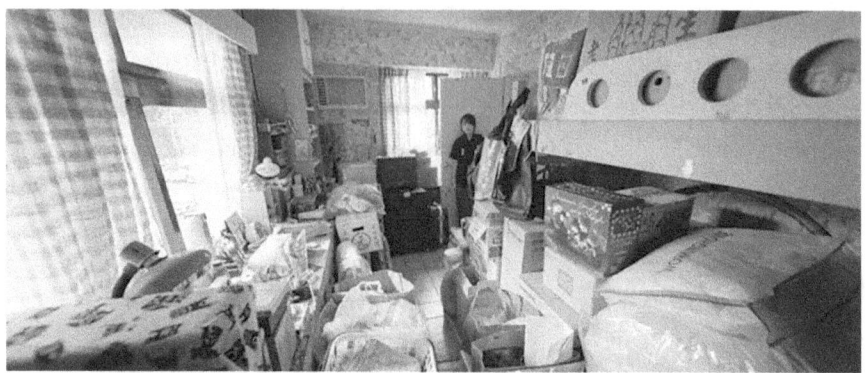

Figure 2: Ruo-min screams in exasperation when she finds her room covered in the family's clutter.[25]
Source: Screenshot from the Netflix drama series *Mom, Don't Do That!* (媽,別鬧了!, 2022), episode 1

forms of nostalgia to be exhibited and revisited regularly throughout the television series, whether in the form of the household's memorabilia or past experiences that they miss. While the characters of *MDDT* do not ever openly comment on the role of the television in the household, the on-screen placement of the traditional television set in the Chen family living room is strategic and represents the gradual obsolescence of the shared, domestic media-viewing experience, as it is affected by SVOD platforms and other personalized, small-screen entertainment.

While Reed Hastings, the cofounder of Netflix, has proclaimed that they offer a "decentralized network" that "break[s] down barriers [so] the world's best storytellers can reach audiences all over the world,"[26] Netflix's decentralizing impact on the domestic media-viewing experience encourages the putting-up, rather than the breaking-down, of barriers. While it

25. An almost identical scenario plays out again in episode 4 when Ruo-min returns to the house again after breaking up with her boyfriend once more.
26. CES, "Reed Hastings, Netflix—Keynote 2016," YouTube, January 6, 2016, 42:39, https://www.youtube.com/watch?v=l5R3E6jsICA.

may be true that Netflix enables narratives to circulate beyond immediate geographical and cultural contexts, these platforms and the media consumption practices they encourage have also redrawn domestic and interpersonal borders along post-TV lines, a phenomenon that *MDDT* self-reflexively draws attention to. *MDDT* depicts the evolving role of the television screen in the domestic space and how this, in turn, alters familial interactions and interpersonal relationships. The gradual obsolescence of the domestic, shared television screen and its decentralized role in the household is self-reflexively portrayed throughout the duration of *MDDT*. In the Chen household, this television set is situated in the living room and is framed by a studio family portrait and a clock[27] while the top of the television set itself is covered in commonplace commemorative objects. As is evident in Ruo-min's return to the Chen household described earlier, the domestic space of the Chen household is filled with nostalgic clutter and like the symbolic, commemorative objects that surround the television set, the domestic, shared television screen of the Chen household has been relegated to the space of nostalgic clutter.

Rather than the all-too-familiar image of a family gathered around the television screen at the end of the day, the television screen of the Chen household is barely used throughout the series, and the family members spend most of their time relying on their personal desktops, laptops, and phone screens for entertainment. Ruo-min even returns home one day to find Mei-mei asleep on the couch with the television playing in the background,[28] a heart-wrenching portrait of domestic loneliness (see Figure 3).

Almost decorative, the domestic shared-screen's decentralization and displacement in the living room is reiteratively represented by the camera

---

27. It is possible to catch glimpses of this domestic, shared, central television screen in the background throughout the drama series, but this television screen is only featured more prominently in one or two scenes, which I will elaborate on later in this article.
28. *Mom, Don't Do That!*, episode 1, performed by Chia-Yen Kuo and Billie Wang.

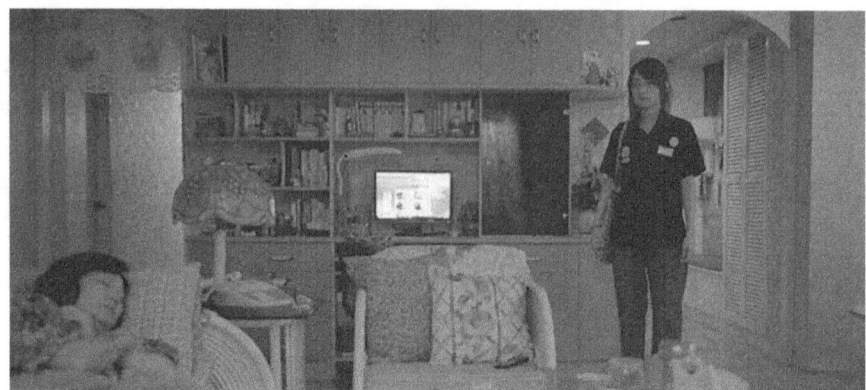

Figure 3: A panning shot from right to left shows an exhausted Ruo-min entering the house to see the television screen on. As she fully enters the living room, she stops in her steps when she realizes Mei-mei has fallen asleep on the couch. The panning shot literally "cuts out" the television screen to show that the monitor of the living room's desktop computer is still on and shows the last things Mei-mei was browsing.[29]
Source: Screenshot from the Netflix drama series *Mom, Don't Do That!* (媽,別鬧了!, 2022), episode 1

angles throughout *MDDT*, which frequently privilege Mei-mei's desktop computer, located to the left of the domestic shared screen (see Figure 4).

When a wheelchair-bound Mei-mei makes numerous online impulse purchases while recovering from an injury, her purchases notably include a new Tatung flatscreen LED display television screen.[30] This new purchase seems to briefly revitalize the domestic shared screen's role in the household and Ruo-min and Cha are seen excitedly trying out the new television set.[31] This revived enthusiasm is short-lived, and the symbolic death of the shared, domestic television screen is further represented when Mei-mei is

---

29. *Mom, Don't Do That!*, episode 1, performed by Chia-Yen Kuo and Billie Wang.
30. Tatung is a multinational appliance company based in Taipei.
31. *Mom, Don't Do That!*, episode 3, performed by Billie Wang, Alyssa Chia, Chia-Yen Kuo, Po-Hung Lin.

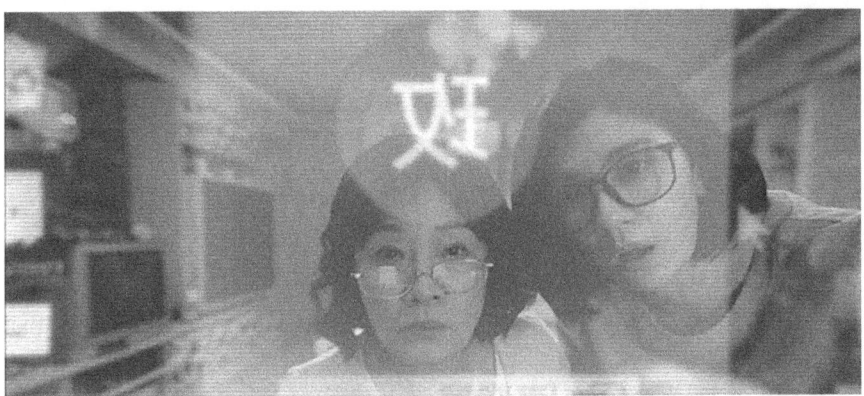

Figure 4: Ru-rong guides Mei-mei on how to set up her online profile. This shot, seemingly taken from inside the screen of the computer, reorients the domestic structure of the Chen family and establishes the personal computer screen as the center of the household. The unused television set is literally and metaphorically relegated to the background.[32]
Source: Screenshot from the Netflix drama series *Mom, Don't Do That!* (媽,別鬧了!, 2022), episode 1

subsequently forced by Ru-rong to return all her impulse purchases, including the newly bought television. Of all the purchases, it is this new flatscreen television that viewers witness Mei-mei having trouble parting with.[33] A superficial assessment of Mei-mei's reluctance might lead to the conclusion that she is merely unwilling to give in to her daughter's interference in her purchases and unwilling to relinquish what little matriarchal and maternal control she has over her household. Yet, Mei-mei's reluctance to part with the television set also represents a more profound and deeper refusal of the symbolic shift and transition from one viewing culture to the next.

The removal of the television set from its central position in the living room radically alters the orientation of domestic space. Even Mei-mei's use

---

32. *Mom, Don't Do That!*, episode 1, performed by Billie Wang and Alyssa Chia.
33. *Mom, Don't Do That!*, episode 3, performed by Billie Wang.

of the computer in the shared living room is short-lived, and when Mei-mei is wheelchair-bound, she starts to use the computer located in her bedroom rather than the one in the living room.[34] In the words of Ramon Lobato, "[The] internet distribution of television content changes the fundamental logics through which television travels, introducing new mobilities and immobilities into the system, adding another layer to the existing palimpsest of broadcast, cable, and satellite distribution."[35] Although Mei-mei's temporary immobility is what prompts her to make this shift, the choice to depict this transition is also a symbolic representation of the ways personal, small screens have altered patterns of mobility and interactions within the domestic space. Mei-mei's physical immobility compels her to try out the new mobility of online shopping, online chatting, and online and overseas dating, all from the convenience of her own bedroom. Even after she recovers, Mei-mei's computer use happens exclusively in her bedroom, symbolizing her complete transition from predominantly occupying the shared domestic living space as the mother of the household to fashioning the private and personal space of her boudoir.[36] In the imagined community of elderly online dating, Mei-mei is free to pursue the new love life she wants, yet this shift also alters the domestic and interpersonal boundaries and interactions of the Chen household. The effects of Netflix and other SVOD and OTT platforms and how they have radically decentralized domestic media-viewing and redrawn domestic and interpersonal borders become apparent.

While the domestic, shared television screen used to demarcate the media-viewing gathering site for a household, the removal or obsolescence of the television decentralizes, disrupts, and disperses social and

---

34. *Mom, Don't Do That!*, episode 3, performed by Billie Wang.

35. Ramon Lobato, *Netflix Nations: The Geography of Digital Distribution* (New York: New York University Press, 2019), 5.

36. Mei-mei puts in a lot of effort to thematically coordinate her bedroom, clothing, and food she prepares with the nationality, culture, and preferences of her foreign suitors, setting up plants, Godzilla figurines, and national flags. *Mom, Don't Do That!*, episode 3, performed by Billie Wang.

domestic-viewing practices, redistributing media onto the various personal electronic devices of our characters. This dispersion, domestic reorientation, scattering, and redistribution of media-viewing centers of the household onto a bunch of smaller personal screens represent a shift from social and domestic viewing practices to more individual and private media-viewing habits. In the wake of Guang-hui's passing, the Chen household must contend with the inevitable divergences in their respective lives and entertainment preferences. While Mei-mei is in her room chatting-up potential suitors online, Ru-rong is at the neighborhood convenience store working on her next novel on her laptop, and Ruo-min is in her room playing computer games.[37] Like the sudden passing of Guang-hui, which the narrative trajectory and soundtrack of *MDDT*[38] encourages us to accept as an inevitable rite of passage that is no less heartbreaking as it is necessary, viewers are invited to consider the symbolic "death" of the domestic television screen as Guang-hui's machine-parallel. Like the accommodating and tireless husband and father-figure that the Chen family finds difficult to move on from, the "passing" of the domestic television screen is an inevitable technological and entertainment rite of passage, a life-altering change we may grieve but must eventually come to terms with and move on from.

---

37. These scenes depicting members of the Chen family household using their electronic devices separately from one another recur throughout the television series. There are no specific episodes or scenes that stand out, but over the course of the television series, viewers learn that what the three women use their devices for and where and how they use these devices are completely different. These differences in their lifestyles and preferences also implicitly explain why they might not watch television together in the living room after Guang-hui's death.

38. The ending song of *MDDT* is Eric Chou's "Graduation" (最後一堂課), which may also be literally translated as "The Last Lesson." Metaphorically referring to the inevitable experience of death, loss, and mortality, the song and its lyrics are thematically tied to Chen Guang-hui's death and the family's struggle to move on. In episode 6, after Chen Guang-hui dies, a brief scene features the gate of the school where Ru-rong teaches, further connecting Chen Guang-hui's death to the song's suggestion of death being the last lesson we all have to reckon with.

Ru-rong does eventually cave and viewers see her shopping online for a new flatscreen television for the family, carefully comparing the prices and specifications.[39] However, in what is arguably the most peculiarly self-referential and metafictive scene that immediately follows the scene of Ru-rong making this new purchase, viewers see the three women sitting on the couch snacking happily while watching television together. When the camera turns to the screen to show us what they are watching and laughing about, the screen is self-referentially screening one of the first establishing scenes from *MDDT* when Ru-rong is arguing with the wife of one of her mother's earlier suitors.[40] At once spectral and specular, this new and surreal television screen in the Chen household projects the television series viewers are watching onto itself and within itself. This nested narrative effect draws viewers into closer proximity to the characters[41] while also inviting viewers to reflect on their own evolving relationships with screens—both big and small, both shared and personal, and both old and new—and contemplate the corresponding changes in the domestic and social relations that are fostered around our media-viewing habits. Apart from the social and interpersonal changes that occur because of this media redistribution from one central and shared domestic screen to many small and personal screens, *MDDT* also invites viewers to contemplate how our media consumption habits and cultures have come to shape other aspects of our domestic and social life, specifically our food-consumption culture. In the same way that *MDDT* seeks to remind its authorial audience of the golden era of Taiwanese entertainment before K-drama's disruption of its regional popularity, *MDDT* evokes transnational nostalgia by subtly relating its depiction of food, foodways, and food consumption to those typical of K-dramas. To understand how *MDDT* relates to K-drama depictions of food and food-consumption

---

39. *Mom, Don't Do That!*, episode 4, performed by Billie Wang, Alyssa Chia, and Chia-Yen Kuo.
40. *Mom, Don't Do That!*, episode 1, performed by Tien Wa and Alyssa Chia.
41. Viewers are momentarily invited by the camera's angle to join the Chen family on their couch to watch the same television series we are already watching.

culture, it is important to contextualize and acknowledge the subtle shifts in the way K-dramas depict food and foodways.

What and How We Eat On-Screen: Media Consumption and Food Consumption Cultures

Media representations of food, foodways, and food-consumption culture have been an increasing interest among both food studies and media scholars. One contributing factor to this interest is the prevalence of food imagery in K-dramas, which has positively contributed to the global demand for Korean food exports.[42] Internationally popular K-dramas like *My Love from Another Star* (별에서 온 그대),[43] for instance, has been credited for popularizing Korean fried chicken, which is widely regarded as the new KFC that has superseded the popularity of the original Kentucky Fried Chicken.[44] As early as 2003, Korean dramas like *Jewel in the Palace* (대장금)[45] were already showcasing and dramatizing traditional Korean cuisine, emphasizing advanced culinary skill, culture, and aesthetics to its East and Southeast Asian viewers—the regional audience that responded well to the initial K-Wave. This media phenomenon is widely recognized among K-drama fanatics, many of whom are actively involved in online forums or discussions about food featured in K-dramas.[46] Evidently, the relationship

---

42. Eunice Ying Ci Lim and Kai Khiun Liew, "Her Hunger Knows No Bounds: Female-Food Relationships in Korean Dramas," in *Routledge Handbook of Food in Asia*, ed. Cecilia Leong-Salobir (New York: Routledge, 2019), 176–92.

43. Tae-yoo Jang, *My Love from Another Star*, performed by Ji-hyun Jun and Soo-hyun Kim, SBS TV, 2013–14, Television Series.

44. This phenomenon is also sometimes referred to as the "*Chimaek* fever," which is an abbreviation of the Korean words for fried chicken (*chikin*; 치킨) and beer (*maekju*; 맥주); Seung-ah Lee, "Soap Opera Drums up Chimaek Fever," Korea.net, March 3, 2014, https://www.korea.net/NewsFocus/Culture/view?articleId=117903.

45. Byung-hoon Lee, *Jewel in the Palace*, performed by Young-ae Lee and Jin-hee Ji, Munhwa Broadcasting Corporation, 2003–4, Television Series.

46. Senater, "An Introduction to K-drama Food," Tell It, September 30, 2021, https://tellit.io/forums/topic/AN-INTRODUCTION-TO-KDRAMA-FOOD/697.

between South Korea's food industry and its entertainment industry is a mutually supportive one, and while viewers tend to dislike conspicuous product placements, K-drama fans generally seem unperturbed and even actively seek out the products that are featured in their favorite dramas.

With fans dedicating so much time and effort to seek out the products they see depicted in their favorite K-dramas, the South Korean entertainment industry has begun to reflect on, tap into, and further develop this phenomenon in their recent dramas. Specifically, they have begun to unapologetically and boldly include tongue-in-cheek, self-reflexive gestures that comically and metafictively call attention to K-drama's influential role in promoting and influencing global food consumption and food-consumption cultures, especially when it comes to the promotion of South Korean food and drinks. Rather than attempt to make their product placements more inconspicuous, K-dramas openly call attention to their many brands and product placements, even boldly featuring these brands and products in the television series that the K-drama characters themselves are watching.

In the recent K-drama *Hi Bye, Mama!* (하이바이, 마마!),[47] Yu-ri, the female protagonist, is seen watching a fictional prime-time Korean television drama called *Her Kimchi Water* (그녀의 물김치) while enjoying cans of beer with her dinner. On Yu-ri's screen, a man and woman are depicted enjoying a home-cooked meal, and viewers of *Hi Bye, Mama!* are thus treated to two instances of on-screen food consumption: Yu-ri eating her dinner with beer and the home-cooked meal that the characters are eating on Yu-ri's television screen. The woman in *Her Kimchi Water* suddenly slams her pair of chopsticks down on the table and asks the male character if the food tastes a little too familiar to him. She then stands, delivers a loud slap to his face, and throws a big red tub of kimchi water at him. Hooked by this rapidly unfolding melodrama on her television screen, Yu-ri's viewing pleasures are

---

47. Je-won Yoo, *Hi Bye, Mama!*, episode 3, performed by Tae-hee Kim, Kyu-hyung Lee, and Bo-gyeol Go, tvN and Netflix, 2020, Television Series.

unfortunately frustrated when her husband's new wife, Min-jung, switches the channel to the news broadcast instead.[48] A subsequent drawn-out montage shows Yu-ri enjoying finger foods like fried chicken with beer while watching K-dramas. Viewers quickly realize that Yu-ri is reminiscing about the many times she enjoyed delicious food while watching her favorite dramas, an activity she misses now that she can no longer partake in such activities as a ghost roaming the mortal realm.

Another example of this is a running gag in the K-drama *Business Proposal* (사내맞선)[49] involving Do-goo Kang, the grandfather of the male protagonist and a semiretired chairman of the Go Food company—an obvious reference to CJ CheilJedang, the prominent South Korean food company best known for the brand Bibigo. Kang, like Yu-ri, enjoys watching a fictional prime-time K-drama called *Be Strong, Geum-hui* (굳세어라 금희야)—another obvious reference to an earlier popular K-drama *Be Strong, Geum-soon!* (굳세어라 금순아)[50]—and many scenes in *Business Proposal* depict Kang watching this K-drama in his pajamas. In one scene, characters of *Be Strong, Geum-hui* have gathered in a pork cutlet shop to discuss the failing marriage proposal between the male protagonist and his wealthy potential in-laws—a common melodramatic scenario and dramatic trope. After the male protagonist confesses that he loves someone else—predictably a woman from much humbler origins and a more modest family background—his offended

---

48. The premise of *Hi Bye, Mama!* focuses on how Yu-ri, a mother who died in a tragic accident, finds herself a ghost roaming the mortal world. Even though her husband subsequently remarried, Yu-ri continues to linger around the household so that she can watch over her family. Yu-ri, as a ghost, frequently fantasizes about the human activities she misses, and one such fantasy has to do with indulging in *chimaek* as she watches her television dramas. The montage in this scene is meant to show us how much Yu-ri enjoyed indulging in food and drink while watching television dramas on her own, so much so that even in death, this is an activity that she craves and recalls fondly.

49. Seon-ho Park, *Business Proposal*, episode 2, performed by Hyo-seop Ahn, Se-jeong Kim, Min-kyu Kim, and In-ah Seol, SBS TV, International, and Netflix, 2022, Television Series, https://www.netflix.com/watch/81509457.

50. Dae-young Lee, *Be Strong, Geum-soon!*, performed by Hye-jin Han and Ji-hwan Kang, Munhwa Broadcasting Corporation, 2005, Television Series.

potential mother-in-law criticizes him for humiliating their daughter and slaps him across the face with a crispy pork cutlet from the table.[51] This absurdly melodramatic exchange witnesses Kang covering his face in disapproval. Later in the same episode, he even self-reflexively comments on how these dramas always depict the wealthy as the drama's villains, and he wonders if people really think all wealthy people behave this badly.

These scenes from recent K-dramas like *Hi Bye, Mama!* and *Business Proposal*, both notably available to watch on Netflix, signify an important but subtle shift in the way K-dramas depict food and food consumption. Bibigo's logo, company, and food products are still prominently featured in many of *Business Proposal's* episodes, but the focus is less on the eating of these food products themselves and more on the company's corporate identity and ethos. The food products prominently featured and eaten in these scenes—fried chicken and canned beer in *Hi Bye, Mama!* and pork cutlet in *Business Proposal*—belong instead to fictional or unnamed brands and establishments. The beer that Yu-ri takes from her fridge all have their brand names facing the front, as if they belong to a strategic product placement, but "Clauster Beer," supposedly a German beer brand, and "Wide Beer," supposedly a brand producing grapefruit beer, are both fictional brands. Rather than feature real brands, these fictional brands promote the food item and influence when and how people consume such food. In other words, the Korean entertainment industry strategically promotes specific food preferences like *chimaek*—also known as the fried chicken and beer fever—and food-related consumer habits (e.g., indulging in *chimaek* while watching K-dramas). Consistently including scenes that depict characters of K-dramas watching K-dramas not only normalizes the watching of K-dramas, as if it is now an everyday activity that even K-dramas themselves cannot avoid depicting if they want to seem realistic, but also allows for self-reflexive meditations on filmic mediations themselves. These scenes

---

51. *Business Proposal*, episode 2, performed by Deok-hwa Lee.

depict a fictional televisual mediating of reality within the already mediated reality of the actual K-drama viewers are watching, creating a nested narrative structure of a television drama within a television drama.

Contrary to what one might assume to be a risky self-reflexive moment that might disrupt the viewers' immersion in the drama's reality and cause them to reflect on how these dramas discreetly seek to influence their consumer behavior, viewers seem to buy into such mediated and self-reflexive portrayals. Netizens on popular online South Korean entertainment forums like theqoo[52] and dcinside[53] started online threads that specifically mentioned the *Her Kimchi Water* scene in *Hi Bye, Mama!*, with a significant number of fellow netizens finding these forum threads and leaving comments about how funny that scene was.[54] One of the anonymous netizens even commented that they immediately noticed the familiar Lock & Lock container[55] for kimchi on the table, and it came to their mind that they would personally have poured out the kimchi water rather than have it continue sitting on the dining table. This thought made the subsequent throwing of the kimchi water on the man even funnier. These forum threads suggest that these self-reflexive scenes not only invite viewers to become more invested in the primary K-drama they are currently watching but also invite viewers to be invested in the secondary K-drama

---

52. "Hi Bye Ma Her Kimchi Water Meme [하바마 그녀의 물김치 짤]," theqoo, February 29, 2020, https://theqoo.net/dyb/1333568597.

53. "Hi Bye, Mama! Gallery: [Normal] Her Kimchi Water [하이바이, 마마! 갤러리: [일반] 그녀의 물김치]," dcinside, February 29, 2020,: https://gall.dcinside.com/board/view/?id=hibyemama&no=739.

54. The thread from theqoo has 1,065 views and two comments, while the thread from dcinside has 648 views and five comments, as of December 9, 2022.

55. Lock & Lock is a household products company based in South Korea. Such containers are commonly used in South Korean households to store kimchi and other food items, and it is my belief that the strategically brief and discreet featuring of a Lock & Lock container in a fictitious television series within another actual television series enables more immersive K-drama world-building. Viewers are invited to share their characters' reality and delight in the pseudorealism (or hyperrealism) of recognizing a product placement that their beloved K-drama characters are likewise not exempt from seeing and being influenced by.

that the characters of the primary K-drama are watching. K-dramas have arguably paved the way for purposeful and economically productive on-screen representations of food in Asian dramas and have reaped the rewards of influencing food-consumption culture through their entertainment offerings. The popularity of these recent Easter eggs and their self-reflexive commentaries on brand endorsement, product placements, and media influence represent a significant shift in how K-dramas innovatively integrate product placements and advertisements into the narrative. These self-reflexive strategies and shifts in product-placement norms in K-dramas, in turn, influence how *MDDT* chooses to depict Taiwanese food and food-consumption culture.

While *MDDT* briefly features a few food and beverage brands throughout the series, such as the brief, recurring scenes in which Ruo-min works as a store manager at TKK Fried Chicken (頂呱呱)[56] (a Taiwanese chain of fried chicken restaurants), the food at these stores are never prominently featured on-screen. Instead, two recurring food-related scenes in *MDDT* feature domestic food preparation and a relatively nondescript roadside food stall, foregrounding family-oriented, home-cooked food and centering local, small businesses instead of recognizable brands and large businesses. The late Guang-hui is known for his culinary expertise, and scenes of the family gathering around the table to enjoy his signature braised pork belly[57] serve as wistful reminders of the domestic dinner table and the warmth of home-cooked meals. With the widow Mei-mei struggling to match her husband's cooking prowess, the two daughters frequently complain about their mother's strange and unpalatable culinary "experiments."[58] Another recurring food-related scene takes place at a thirty-five-year-old roadside food

---

56. *Mom, Don't Do That!*, episode 3, performed by Chia-Yen Kuo and Chloe Xiang.

57. *Mom, Don't Do That!*, episode 6, performed by Billie Wang, Alyssa Chia, Chia-Yen Kuo, and Johnny Kou.

58. *Mom, Don't Do That!*, episode 2, performed by Billie Wang, Alyssa Chia, Chia-Yen Kuo, and Johnny Kou.

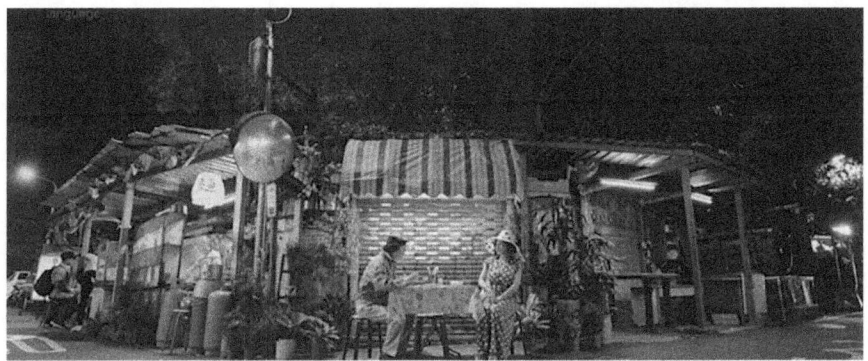

Figure 5: This is the first instance of Zuo's roadside stall being featured in *MDDT*. On the left, Zuo is preparing the food at his store and on the right, Mei-mei and Yong, one of her early suitors, are seated in front of the shutters with yellow poppies, having a conversation about their relationship over food.[59]
Source: Screenshot from the Netflix drama series *Mom, Don't Do That!* (媽,別鬧了!, 2022), episode 1

stall (see Figure 5), affectionately known only as Mr. Zuo's Old Stall and referred to by the characters as "the old place" (老地方).[60]

The Chen family is shown to frequent the stall, even before the passing of their Guang-hui, so much so that Zuo,[61] the owner of this roadside food stall, knows them and knows their standard order of braised meat, dry noodles, and wonton soup by heart. *MDDT* romanticizes this food stall, indulging its viewers in the elaborate fantasy of nostalgic regularity,

---

59. *Mom, Don't Do That!*, episode 1, performed by Billie Wang and Shao-Hua Lung.
60. The expression "old place" (老地方) is frequently used in Chinese-speaking communities to refer to a usual and preferred meeting place. The expression does not necessarily mean that the place is old and has a positive connotation that generally suggests it is a place that people frequent and are fond of.
61. The ladies of the Chen family address the owner of this roadside stall as 左伯伯 (Zuo bêh4 bêh4; Uncle Zuo) and 老左 (Old Zuo), which signify their regular patronage and familiarity with the owner.

familiarity, domestic ritual, and impossible permanence. No matter which combination of characters patronizes this stall and no matter the time, they are always seated at the same table with the backlit painting of yellow poppies on the shutters behind them. No matter how often the interpersonal disputes and whims of the Chen family disrupt the dining experience of other patrons—Mei-mei takes a dish from another table without their permission to show some foreigners what they should order[62]—the other patrons are always kind enough to shrug these inconveniences off without kicking up a fuss. This on-screen portrayal of food, foodways, and food-consumption culture is at once distinct from and a part of the K-drama model.

MDDT implicitly suggests that the K-drama model of food consumption favors eating out, ordering in, eating alone, and bigger, well-established brands. Although it might appear as if *MDDT* is rejecting the K-drama model of representing food on-screen, these depictions of home-cooked food and roadside stalls still serve to promote Taiwanese food and local street-food consumption. Through the many recurring and romanticized depictions of home-cooked meals and Zuo's roadside stall, *MDDT* reminds viewers of Taiwanese dramas and how their local food cultures used to dominate many of the screens in East and Southeast Asia, presenting yet another nostalgic reminder of a bygone era of regionally popular Taiwanese entertainment. Adopting and taking a leaf out of K-drama's playbook, *MDDT* strategically imbues local dishes and food cultures with media, cultural, and affective significance. Since credit scenes on Netflix are often cut short by the automatic transition into the "Next Episode" or the automatic playing of trailers from Netflix's entertainment recommendations, the format of the platform enables a strategic and convenient overlooking of the credits. This overlooking of the credits facilitates the discretion of the impressive range and number of sponsors and brands backing the production of this

---

62. *Mom, Don't Do That!*, episode 7, performed by Billie Wang.

Taiwanese drama.[63] Thus, *MDDT*'s response to the K-drama model of mediated food representation acknowledges the influence of this K-drama model while simultaneously drawing attention to how this model has altered viewers' food consumption habits and preferences in ways that they might not have been aware of, such as in its subtle critique of mukbang.

MDDT critiques the K-drama model of representing food and food consumption culture in its depictions of the mukbang (먹방, eating broadcast) phenomenon[64] and masculine cooking[65]—the former being a social media phenomenon that originated in South Korea and the latter being a well-established K-drama trope. Dashi Yaoji (大食妖姬)—the online

---

63. By clicking on the "Watch Credits" option or toggling the Autoplay Next Episode setting on the platform, viewers of *MDDT* will gain access to the list of Taiwanese and international sponsors and supporting organizations for the television drama, which range from the Taiwanese Uni-President Enterprises Corporation (統一企業公司, the company responsible for Tung-I instant noodles), the Taipei Culture Foundation (台北市文化基金會), the Berlin-based but Asia's leading online food and grocery delivering platform foodpanda, and South Korean multinational electronics corporation Samsung.

64. For studies that problematize the mukbang phenomenon and its negative effects, see Kang, Lee, Kim, and Yun (2020); Kircaburun, Yurdagül, Kuss, Emirtekin, and Griffiths (2020); and Jeon and Ji (2021). For studies that delve into the motivations, desires, and sociality involved in the mukbang phenomenon, see Spence, Mancini, and Huisman (2019); Kim (2021); and Choe (2019); Eun-kyo Kang, Ji-hye Lee, Kyae-hyung Kim, and Young-ho Yun, "The Popularity of Eating Broadcast: Content Analysis of 'Mukbang' YouTube Videos, Media Coverage, and the Health Impact of 'Mukbang' on Public," *Health Informatics Journal* 26, no. 3 (2020): 2237–48; Kagan Kircaburun, Cemil Yurdagül, Daria Kuss, Emrah Emirtekin, and Mark D. Griffiths, "Problematic Mukbang Watching and Its Relationship to Disordered Eating and Internet Addiction: A Pilot Study among Emerging Adult Mukbang Watchers," *International Journal of Mental Health and Addiction* 19, (2021): 2160–69; Chang-young Jeon and Yunho Ji, "A Study on Irrational Consumption Tendency According to Exposure of Video Contents of Mukbang (Eating Broadcasts) and Cookbang (Cooking Broadcasts)," *International Journal of Tourism Management and Science* 36, no. 1 (2021): 23–40; Charles Spence, Maurizio Mancini, and Gijs Huisman, "Digital Commensality: Eating and Drinking in the Company of Technology," *Frontiers in Psychology* 10 (2019): 1–16; Yeran Kim, "Eating as a Transgression: Multisensorial Performativity in the Carnal Videos of *Mukbang* (Eating Shows)," *International Journal of Cultural Studies* 24, no. 1 (2021): 107–22; Hanwool Choe, "Eating Together Multimodally: Collaborative Eating in *Mukbang*, a Korean Livestream of Eating," *Language in Society* 48, no. 2 (2019): 171–208.

65. Jooyeon Rhee, "Gender Politics in Food Escape: Korean Masculinity in TV Cooking Shows in South Korea," *Journal of Popular Film and Television* 47, no. 1 (2019): 56–64.

handle of the young lady that vies with Ruo-min for Cha's affection—is a popular livestreamer who films herself eating copious amounts of food. Like the many South Korean mukbang broadcast jockeys, Yaoji is pretty, young, and feminine, adopting a sweet and bubbly persona.[66] Portraying her as a somewhat endearing villain who tricks Ruo-min into sponsoring her with TKK Fried Chicken for one of her livestream sessions, Yaoji's on-screen personality belies her potential for vindictiveness and petty vengeance. If the home-cooked meals of the Chen family[67] and Zuo's roadside stall represent forgotten and underappreciated Taiwanese food cultures and sociality, the deluge of scathing, anonymous online comments that Yaoji's mukbang livestreams receive is the cultural antithesis. Even as an angered Ruo-min spitefully forces Yaoji's face into a pile of chicken wings and fries and pours a bottle of coke over her head on the livestream, Yaoji's followers continue to leave comments, variously delighting in the spectacle, mocking Yaoji, and cynically asking if the conflict is staged.[68] While K-dramas frequently have embedded mukbang-like sequences, such as the earlier scene described in *Hi Bye, Mama!* when Yu-ri is seen indulging in *chimaek* as she watches her K-dramas, these scenes are meant to induce food cravings and are not meant to critically reflect on the implications and effects of mukbang. By inserting a character like Yaoji as an on-screen embodiment of this phenomenon, *MDDT* foregrounds the negative effects of an individualistic, materialistic, overindulgent, and narcissistic food-consumption and media-consumption culture,[69] problematizing how K-Wave content and OTT media-consumption culture have coproduced these negative outcomes. By the end of *MDDT*, Yaoji has given up on chasing Cha and is preparing to further her studies to become a primary school teacher,[70] a

---

66. *Mom, Don't Do That!*, episode 4, performed by Chloe Xiang.
67. *Mom, Don't Do That!*, episode 1, performed by Billie Wang, Johnny Kou, Alyssa Chia, Chia-Yen Kuo, and Kang-ren Wu.
68. *Mom, Don't Do That!*, episode 4, performed by Chia-Yen Kuo and Chloe Xiang.
69. *Mom, Don't Do That!*, episode 7, performed by Chloe Xiang.
70. *Mom, Don't Do That!*, episode 7, performed by Chloe Xiang and Chia-Yen Kuo.

fate that symbolizes the rehabilitative food-media relationship that *MDDT* indirectly offers to its viewers.

Reflecting on yet another K-drama trope of masculine cooking, *MDDT*'s characterization of the character Senior—the supposed love interest of Ru-rong—is yet another attempt to criticize the gendered media-food relationship that K-dramas have normalized. The K-drama trope of masculine cooking often features an attractive and financially successful man rolling up his sleeves to whip up delicious and profession-ally plated food that is then served to the lucky woman he is attracted to.[71] Senior is an over-the-top representation of this K-drama archetype. Shortly after he runs into Ru-rong at the convenience store, Senior looks deeply into her eyes, shares that he is good at cooking, and invites her to his place so that he may prepare a meal for her.[72] A smitten Ru-rong will eventually be seen visiting Senior's home and delighting at the perfectly portioned and professional-looking meals he prepares for her.[73] Senior, with his stylish bachelor pad, successful career, substantial wealth, and professional cooking standards[74] checks all the boxes of a stereotypical male lead in a K-drama. The drama makes it abundantly clear that the character is meant to be interpreted in relation to this K-drama trope and archetype, and a Korean song plays in the background during a melo-dramatic rainfall scene between Ru-rong and Senior. Senior runs out in the rain to meet Ru-rong and the slightly off-key Korean song plays in the background as the usually glib Senior uncharacteristically stutters

---

71. This is yet another K-drama trope that is not often acknowledged among scholarly com-munities but is well-established among K-drama fans. The fan-made YouTube video "6 K-Drama Scenes that Prove We Want Our Men to Cook for Us," with twelve thousand views, compiles six different K-drama montages that feature an attractive male character cooking. The video demonstrates how common these scenes of masculine cooking are in K-dramas and how aesthetically pleasing these scenes are meant to be. Amusing Sphere, "6 K-Drama Scenes that Prove We Want Our Men to Cook for Us," YouTube, Febru-ary 9, 2021, 4:15, https://www.youtube.com/watch?v=ciszKngW8Rw.

72. *Mom, Don't Do That!*, episode 1, performed by Alyssa Chia and Kang-ren Wu.

73. *Mom, Don't Do That!*, episode 3, performed by Alyssa Chia and Kang-ren Wu.

74. *Mom, Don't Do That!*, episode 3, performed by Alyssa Chia and Kang-ren Wu.

over what to say to Ru-rong.[75] This parodic Korean song titled "Want Rock-Hard Pecs" (我的胸很硬) is composed and sung by Taiwanese lyricist Matthew Yen (嚴云農), who refers to this song as "the perverse Korean song from episode 7" (第七集的變態韓國歌) in an Instagram post promoting the *MDDT* television series.[76] These metareflexive and self-referential commentaries in *MDDT* are akin to a long trail of bread-crumbs throughout the drama, casting doubt on Senior's character and inviting viewers to laugh at how unrealistic and problematic this K-drama archetype is. Senior's qualifications and mysterious charm à la Christian Grey barely disguise his possessiveness, his controlling personality, and his erratic outbursts, and viewers are encouraged to consider the possibility that he might be a psychopath and a serial killer.[77] Only in episode 9 do the viewers learn that Senior has always been a figment of Ru-rong's imagination, is merely a new character in the novel she is writing, and his appearance is inspired by a staff member working at the convenience store that Ru-rong often writes at. Senior, as an exaggerated portrayal of the typical K-drama male lead, becomes a strategic counterstereotype to the masculine domesticity that K-dramas have been centering and encouraging its viewers to fantasize about.

As a Taiwanese drama responding to K-dramas' dominance, *MDDT* establishes this masculine stereotype as not just improbable and unrealistic

75. *Mom, Don't Do That!*, episode 7, performed by Alyssa Chia and Kang-ren Wu.
76. Matthew Yen (嚴云農, kumono_yan), "而我最想給你的，是一雙翅膀，讓你從我的世界，飛出去 . . .," [But what I wish to give to you the most, is a pair of wings, so that you may take flight and depart from my world . . .,] Instagram, July 25, 2022, https://www.instagram.com/p/CgbKwKGB4jf/. It is worth noting that this "Korean love song" is written in grammatically incorrect Korean or broken Korean, and part of the song's lyrics even include comedic lines that can be roughly translated as "I want to sing Korean song, but I don't know Korean," "Let's go Korean," and "Sing Korean song, it doesn't matter if I don't understand it," ironically highlighting how Korean songs, especially those with romantic themes, are popular even though many international consumers of K-dramas do not know what the songs are actually about. The author of this article would like to thank Dr. Na-hyun Kim for her help in translating and transcribing the lyrics of this song.
77. *Mom, Don't Do That!*, episode 9, performed by Alyssa Chia and Kang-ren Wu.

but as laughably banal and inane as it is unhealthy and abusive. Rather than a paragon of progressive masculinity and eligibility, *MDDT* boldly recasts and problematizes the aspirational K-drama male archetype as a dysfunctional, dangerous, and morally depraved person whose traits should not inspire desire but instead serve as cautionary red flags. This huge narrative twist, if unanticipated by the viewer prior to episode 9, proves *MDDT*'s point that K-dramas have normalized this problematic archetype and romantic dynamic to the extent that we no longer even question the nature of this unlikely and untoward relationship. The breadcrumbs—or clues—scattered throughout the television drama's duration are painstakingly apparent. Senior's image has always been discreetly featured on all the covers of Ru-rong's published romance novels.[78] In another self-reflexive moment, Ru-rong dreamily talks about the good looks of Taiwanese actor Kang-ren Wu to convince Ruo-min to break up with her unfaithful boyfriend, naming the actual Taiwanese idol-drama actor who later plays the role of Senior in *MDDT*.[79] When Ru-rong first meets Senior in the convenience store, she has also taken off her glasses and needs to squint to recognize him, a minor detail that is meant to foreshadow the metaphorical blindness of those who fail to recognize these apparent tells that would suggest Senior is not a reliable or real character.[80] Mei-mei even foreshadows this twist when she tells Ru-rong that convenience stores will not provide her with the inspiration she needs to write a romance novel,[81] further hinting at the lack of genuine romantic fulfilment between Ru-rong and

---

78. *Mom, Don't Do That!*, episode 1, performed by Billie Wang, Alyssa Chia, and Kang-ren Wu.

79. *Mom, Don't Do That!*, episode 1, performed by Alyssa Chia, Chia-Yen Kuo, and Kang-ren Wu. I refer to the actor Kang-ren Wu or Chris Wu as a Taiwanese idol-drama actor because one of his most well-known roles is that of the supporting male lead Hua Tuo Ye in the idol-drama *Autumn's Concerto* (2009). Notably, Ru-rong's adulation of actor Kang-ren Wu is also self-reflexive in that Alyssa Chia—the actress who plays Ru-rong—recently starred in another Taiwanese television drama, *The World Between Us* (2019), alongside Kang-ren Wu.

80. *Mom, Don't Do That!*, episode 1, performed by Alyssa Chia and Kang-ren Wu.

81. *Mom, Don't Do That!*, episode 2, performed by Billie Wang and Alyssa Chia.

Senior, whose relationship is ultimately confined to Ru-rong's active writer's imagination.

Conclusion: The Uncertain but Hopeful Horizons of Global Taiwanese Television Drama

By employing these self-reflexive devices as a means of establishing metadramatic continuity between Taiwanese drama-consumption cultures in the 2000s and this new Taiwanese Netflix original, *MDDT* invites its viewers to recognize and acknowledge their "shared sensibility," "shared creative understanding," and "shared worldview or outlook."[82] This shared experience does not exclusively pertain to the domestic, familial, romantic, or even cultural resonances but extends to their collective experience of the transformations in media-viewing habits and the dwindling regional popularity of Taiwanese dramas, as it is impacted by the popularity of K-dramas and changes in media-viewing cultures. Viewers may recognize how their media-consumption habits and preferences are complicit in enabling the normalization of certain dramatic tropes, media-viewing cultures, and media-influenced social relations, yet these media-consumption habits and preferences are simultaneously indispensable in remedying and revising the less desirable effects of such normalization.

The illusion that *MDDT* indulges its implied authorial audience in is one in which the regionally popular Taiwanese dramas of yesteryear, whether Minnan family melodramas or Mandarin idol-dramas, collectively represent a romantic, domestic, and sociocultural phase that viewers may look back on fondly and send off like an old and dear friend or like a beloved family

---

82. W. Michelle Wang, 【這又不是演戲】 'We're not playacting here,'" *JNT: Journal of Narrative Theory* 45, no. 1 (2015): 107–8. These are all expressions that Wang uses to describe the narrative and affective effects of self-reflexive devices on the implied authorial audience of Taiwanese idol-dramas.

member. *MDDT* grants their "long-lost"[83] viewers symbolic closure by met-aphorically depicting the untimely and regrettable "death" of Taiwanese dra-mas, represented by the abrupt deaths of Chen Guang-hui[84] and Mei-mei's lifelong best friend Jin.[85] Although the deaths of the two characters take place five years apart, the two deaths occur one after the other within the anachronistic narrative trajectory of *MDDT*, with the details of Guang-hui's death unfolding in episode 6 and Jin's death and funerary rites unfolding in episode 7. The former dies just as he is planning to bring his wife on a long-anticipated trip around the world, while the latter dies after reconcil-ing with an estranged friend. These untimely deaths come to represent the regrettable and premature "death" of Taiwanese dramas and their growing regional popularity in the 2000s, which was unfortunately stunted by the popularity and predominance of K-dramas. I consider this metaphorical on-screen death and farewell to Taiwanese dramas of the past as an elaborate and beautiful illusion because the regional and global popularity of enter-tainment offerings from South Korea and Taiwan will remain an ongoing contestation and Taiwanese Minnan prime-time family melodramas and Mandarin idol-dramas have not really "died out" at all. In fact, *MDDT*'s production background might even suggest that the relationship between the two entertainment industries may move toward a more inter-Asian, global-Asian,[86] or transnational collaborative model.

Ironically, the symbolic on-screen deaths, farewells, and moving-on from Taiwanese dramas of the past, as they are poignantly depicted in

---

83. "Long-lost" is in scare quotes as these viewers have never really gone anywhere but have simply diversified or changed their media-viewing preferences. Some viewers, like the au-thor of this article, have never really left and are still avid supporters of Taiwanese dramas.

84. Although viewers know since the start of the series that Chen Guang-hui died, the twelve hours that lead to Chen Guang-hui's death are only covered in detail in episode 6 of the television series.

85. Jin is Mei-mei's life-long best friend. Episode 7 of the television series focuses on her passing and funerary rites.

86. Depending on one's preferences, this phenomenon may also be described as or belong to existing Pan-Asian or Trans-Asian discourses.

MDDT, potentially usher in a large-scale revival and revitalization of the Taiwanese entertainment industry, and this hope at revitalization is represented by *MDDT*'s significant departures from old Taiwanese dramatic tropes. Rather than the domestic reunions and matrimonial or romantic unions typical of the happily-ever-after endings that characterize Taiwanese dramas of the past, *MDDT* ends with a series of necessary but open-ended departures, dispersions, and separations that are at once hopeful and wistful. After selling their home, Mei-mei marries Robert and relocates to Australia to live among his eclectic trailer community. The two sisters continue to live in Taiwan but now live separately, seemingly content with their newfound independence and solitude. Akin to a compelling yet invisible thread connecting its implied authorial audience, the somewhat ambivalent ending of *MDDT* invites viewers to gather at this shared and mediated sociocultural threshold, where they, too, might consider what their next step might be in terms of their media-viewing preferences, habits, influences, and effects.

Playfully foregrounding the less desirable effects of K-dramas and post-TV SVOD and OTT media consumption cultures and poking fun at how viewers are captivated by absurd and predictable plotlines and archetypes, *MDDT* gives pause to the uncritical consumption of K-dramas, Taiwanese dramas, and SVOD and OTT-hosted media content. With its ambivalent conclusion, *MDDT* gestures AT the emergent and uncertain possibilities and horizons of global Taiwanese television drama. If there is a happy ending or resolution to be found at the end of the series, it is the implicit reassurance and promise that Taiwanese television dramas, like *MDDT*, will continue to evolve and will remain an available and increasingly global entertainment option.

Acknowledgments

The author of this manuscript would like to thank Dr. Nahyun Kim of Drexel University for her help in translating the lyrics of the song "Want Rock-Hard Pecs" from Korean to English.

How *Pachinko* Mirrors Migrant Life

Rethinking the Temporal, Spatial, and Linguistic Dimensions of Migration

WINNIE YANJING WU

Abstract

This paper critically analyzes the Apple TV+ series *Pachinko* (2022) to comprehend its cross-historical and cross-regional metanarrative unfolding from the organization of temporality, spatiality, and language. As the TV adaptation of Min Jin Lee's eponymous novel, *Pachinko* depicts a family's migration journey from Korea to Japan after the 1910s and emphasizes their suffering from systemic discrimination against temporary Korean residents. Produced by talents from Korea, Japan, and the United States, *Pachinko* displays strong hybridization that combines American TV conventions with a distinct East Asian culture and history. The hybridized, multicultural, and multilingual background of the production necessitates a transnational and interdisciplinary framework to analyze its critical success and cultural implications. Expanding Harvey's notion of time-space compression, the paper conceptualizes the temporal and spatial experience of watching a transnational production via global streaming as a mirrored experience of migrant life. It tackles television dramas as a strategy to understand contemporary migration and globalization by first outlining the evolutionary trajectory of television, and then identifying the movements, mobility, and the transnational cultural flows in *Pachinko*. Moreover, this paper analyzes the linguistic aspects of *Pachinko*, particularly in translation and multilingualism, to establish a connection between language and cultural identities. Inquiring into previous literature on translation, this paper also seeks to understand the complexity of communicating in multiple languages, both literally and metaphorically. Finally, this paper examines how migration and migrants are reimagined in *Pachinko* at a time when national borders and cultural and linguistic barriers are quickly eroded by global streaming TV.

Keywords: Transnational Television, Asian American Studies, East Asian Popular Culture, Migration Studies, Pachinko

Introduction: K-drama, American Style?

After the success of several K-dramas and movies in global distribution, such as *Kingdom* (2019), *Parasite* (2019), *Squid Game* (2021), and *Minari* (2021), streaming platforms are creating space for an increasing number of South Korean productions. The premiere of *Pachinko* on Apple TV+ represents the first time that American culture has delved deeply into the painful migration journey of Koreans during the Japanese occupation of Korea. Adapted from the original novel written by Min Jin Lee, *Pachinko* is a family saga that spans eighty years of history from 1910 to 1989, depicting a family's migration from Busan, Korea, to Osaka, Japan. In 1910, The Empire of Japan colonized Korea and attempted to erase any traces of Korean culture and language. In the process of forced Japanese assimilation, many Koreans lost touch with their cultural roots and eventually moved to Japan for better chances of survival. Many of them lived in Japan as temporary residents where they faced systemic discrimination that also extended to their Japan-born descendants. Sunja's family represents the life story of many Korean migrants, and the character of the grandson, Solomon, serves as "the product of a clash of different countries tied to historical animosity."[1] The

---

1. Seung-hun Oh, "'Pachinko'—a Story by and about Korean Diaspora—Captivates the World," Hani, May 2022, https://english.hani.co.kr/arti/english_edition/e_entertainment/1042830.html;

series speaks about the culture and history that are uniquely Korean, such as the story of "Zainichi," temporary residents of Japan since the 1910s, and "han," a term created to capture the sorrow, grief, anger, and regrets that were endured across Korean history. There are also universal themes that resonate with the global audience in general, such as family, memories, and trauma. Besides high expectations from cultural critics, statistic show that in the first two weeks of release, *Pachinko* ranked as the number 1 most watched television show across all over-the-top (OTT) platforms.[2]

In terms of form and style, *Pachinko* is reviewed as a "K-drama in American style."[3] The original novel of *Pachinko* was written in a linear story order, but the adapted TV series intentionally breaks this linearity. The intense crosscutting and constant eclipses and flashbacks disrupt the audience's usual perception of time and space. The creative choice displays apparent traits and conventions of complex TV, an era of TV production in America since the 1990s. Theorized by Jason Mittell, the key feature of complex TV includes continuous narrative enigmas, moments of narrative spectacles, and operational aesthetics.[4] As suggested by Mittell, "Complex narratives often reorder events through flashbacks, retelling past events, repeating story events from multiple perspectives, and jumbling chronologies."[5] Following these conventions, *Pachinko* also incorporates complexity by constantly altering the time lines of multiple stories that deliberately confuse temporality.

Ashley Hajimirsadeghi, "Pachinko Review: Revisiting Korean History and Diaspora," MovieWeb, March 25, 2022, https://movieweb.com/pachinko-review/;

Kia Fatahi, "'Pachinko': Revealing History through the Lens of a Korean Family," *Observer*, April 2022, https://fordhamobserver.com/68804/recent/arts-and-culture/pachinko-revealing-history-through-the-lens-of-a-korean-family/.

2. Geca Wills, "Lee Min Ho, Kim Min Ha's 'Pachinko' Dominate OTT Ranking for Two Consecutive Weeks," KDramaStars, April 12, 2022, https://www.kdramastars.com/articles/124586/20220411/lee-min-ho-kim-ha-pachinko-dominate-ott-ranking.htm.

3. Mike Hale, "'Pachinko' Review: K-Drama, American-Style," *New York Times*, March 24, 2022, https://www.nytimes.com/2022/03/24/arts/television/pachinko-review.html.

4. Jason Mittell, *Complex TV: The Poetics of Contemporary Television Storytelling* (New York: New York University Press, 2015).

5. Mittell, "Complex TV," 36.

Another element that adds to the complexity of *Pachinko* is multilingualism. According to *Time*, *Pachinko* is the largest multilingual show ever in the history of Hollywood.[6] In the process of finding distributing platforms, Soo Hugh, the writer of *Pachinko*, insisted that streaming in native languages is the only way of presenting this story.[7] The switching of languages plays a vital role in character-building, as it reflects cultural identities and personal history. Before the start of each episode, a user-interface menu appears that states, "*Pachinko* is presented in its two original languages, Korean and Japanese. To turn on subtitles or dubbed audio, pause the video and choose the following icons." After, the interface continues to state, "Japanese dialogue subtitles in blue, Korean dialogue subtitles in yellow." The interface requires the audience to select the most appropriate translation for them before the show can proceed playing. The audience cannot idly wait for the show to automatically start playing itself. Thus, apart from narrative complexities, *Pachinko* also demands the audience's capacity to listen to multiple languages and to comprehend multilayered linguistic references. This makes *Pachinko* relatively challenging compared to other domestic-themed melodramas. *Pachinko* requires a great deal of attention to watch, whether it is in reading subtitles, identifying languages being spoken, or keeping up with the story world.

Recruiting resources and talents across three countries, *Pachinko* is a truly hybrid, multicultural, and multilingual production that targets a transnational audience on Apple TV+. The series also makes itself unique from the previous coproductions because the story constantly relocates between Korea, Japan, and the United States. In other words, it is hard to pinpoint the nationality of *Pachinko*, which is different from the traditions of soap

---

6. Andrew Chow, "How Apple TV+ Made 'Pachinko,' One of the Biggest Multilingual Shows Ever," *Time*, March 2022, https://time.com/6157906/pachinko-apple-tv/.

7. "'Joy Luck Club' to 'Pachinko': The Asian Diaspora On-Screen," NPR, May 2022, https://www.npr.org/2022/05/20/1100380098/joy-luck-club-to-pachinko-the-asian-diaspora-on-screen.

operas that often emphasize one specific nationality. When I watched the series, these questions appealed to me: Is *Pachinko* an Americanized K-drama or a globalized American drama? How does *Pachinko* depict a migrant life by constructing temporality and spatiality? Can streaming free us from national borders, time zones, and language barriers? How do we connect multilingualism to conceptual frameworks of personal history and identities? These questions subsequently become the research questions of this essay to investigate the temporal, spatial, and linguistic dimensions of migration. By looking into the perception of time, space, and language in *Pachinko*, the essay seeks to identify the cultural implication of watching transnational programs in the age of digitization and globalization.

Migrant/Migration Melodramas

Mainstream TV reviews generally regard *Pachinko*'s genre as an epic and a family saga, due to its emphasis on genealogy and a long chronological span.[8] However, *Pachinko* is indeed a show about migration, as the family is scattered across three different countries and continues to take a diasporic path. Migration is the drive, motivation, and twist in almost all the characters' fates. Scholars in literary studies have termed these characteristics

---

8. Caryn James, "Pachinko Review: A 'Dazzling, Heartfelt Korean Epic,'" BBC, March 2022, https://www.bbc.com/culture/article/20220323-pachinko-review-a-dazzling-heartfelt-korean-epic.

  Stuart Jeffries, "Pachinko Review—a Sumptuous South Korean Epic like Nothing Else on TV," *Guardian*, March 25, 2022, https://www.theguardian.com/tv-and-radio/2022/mar/25/pachinko-review-min-jin-lee-south-korea.

  John Powers, "Deeply Felt and Unpredictable, 'Pachinko' Follows the Epic Rise of a Korean Family," NPR, March 23, 2022, https://www.npr.org/2022/03/23/1088179222/pachinko-apple-tv-review.

  Daniel D'Addario, "'Pachinko' Is a Multigenerational Saga Whose Time Jumps Sap Its Power: TV Review," Variety, March 24, 2022, https://variety.com/2022/tv/reviews/pachinko-tv-review-1235213057/; Kathryn VanArendonk, "Pachinko Builds Epic Family Drama on an Exquisitely Intimate Scale," Vulture, March 21, 2022, https://www.vulture.com/article/pachinko-review-apple-tv-series.html.

as the genres of migration melodramas and migrant melodramas.[9] Ford's "migration melodrama" refers to Hong Kong's 1980s and 1990s movies like *An Autumn's Tale* (1987), which typically centers on "a range of stories of people who left Hong Kong for new lives elsewhere."[10] The generation of Hong Kong migrants from the 1980s to 1990s are called "astronauts" because they often transit back and forth from Hong Kong, spending a fair amount of time on planes. On the other hand, Puga's definition of migrant melodrama refers to contemporary cultural production that "trains a melodramatic imagination on migrants and emphasizes suffering as a necessary step in the process of inclusion."[11] Her analyzed examples focus on the representation of undocumented child migrants and the underlying logic of the "political economy of suffering" in these works.[12] As *Pachinko* chronicles the eventful period from 1910 to 1989, the family's migration history evolves from suffering migrants, such as Sunja and her peers, to flexible autonomous migrants, such as Sunjia's grandson Solomon in his later life.

As *Pachinko* begins, travel and migration have been intimately associated with tragedies and traumatic memories. Each time a character moves, he or she is either escaping from a catastrophe, a major life crisis, or a devastating loss. In this particular drama, almost all relocations can be considered involuntary and forced migrations because the characters run away from misery, leaving their homeland with profound regrets and uncertainties. Sunja leaves her hometown of Yeongdo for Osaka after ending a shameful and all-consuming affair with the wealthy married man Hansu. In Hansu's formative years, he also leaves the town of Yokohama

9. Ana Elena Puga, "Migrant Melodrama and the Political Economy of Suffering," *Women & Performance: A Journal of Feminist Theory* 26, no. 1 (January 2, 2016): 72–93, https://doi.org/10.1080/0740770x.2016.1183982.

Stacilee Ford, *Mabel Cheung Yuen-Ting's an Autumn's Tale* (Hong Kong: Hong Kong University Press, 2008).

10. Ford, "Mabel," 2008.

11. Puga, "Migrant," 72.

12. Puga, "Migrant," 72.

after losing his home and family in the 1923 Great Kanto Earthquake and then takes on his long-term exile. Solomon, the youngest character in the series, is sent to the United States after being caught in a shoplifting incident and fistfight in Osaka. In the season finale, Solomon decides to face the bygones and unresolved trauma by staying in Osaka and continuing the family business. He refuses to flee to America again by saying, "America is not the solution, it is a fantasy." Solomon's decision marks the end of his family's eighty-year history of escapism, exile, and living in displacement. His decision to return to Osaka suggests the "mobility turn" that is missing from his grandmother's generation—migration is no longer a one-way flow but multiple transnational movements.[13] Solomon's first migration was involuntary—as it was arranged by his father—but his second time was a voluntary and spontaneous choice. Migrants like Solomon demonstrate multiway mobility in their capacity to flow back and forth between places due to factors such as a high-skilled work background (Solomon's educational and professional credentials) and capital accumulation (Solomon's generational family business).

As *Pachinko* emphasizes the history of suffering, it still touches on many hopeful and heartening moments, such as celebrating the birth of a new baby, reuniting with a long-lost friend from the homeland, and completing the dream of a dying lover toward the end of her life. The bright side of an involuntary migration is the chance for a new beginning and the betterment of life. Having overcome the pains of separation and alienation, the migrants usually find more hope than continuous misery. Though suffering and survival set the tone of the migrant narrative in *Pachinko*, the opportunities to start fresh and live differently, due to the stoic and persistent acts of migration, are equally significant.

---

13. Claire S. Lee, "Temporal Dimensions of Transient Migration Studies the Case of Korean Visa Migrants' Media Practices in the US," in *The Routledge Handbook of Digital Media and Globalization*, ed. Dal Yong Jin (New York: Routledge, 2021).

Theoretical Framework: Time-Space Compression and Television Evolution

Media communication has undergone tremendous technological evolution since the 1980s. Media has brought our "imagined" global community to life.[14] Television, especially after satellite TV, has enabled us to experience "a rush of images from different spaces almost simultaneously, collapsing the world's space into a series of images on a television screen."[15] This idea of a shrinking world affected by media and transportation technology is what Harvey terms the "time-space compression."[16] It refers to the fact that "the time horizons of both private and public decision-making have shrunk, while satellite communication and declining transport costs have made it increasingly possible to spread those decisions immediately over an ever wider and variegated space."[17] After the 1990s, some scholars have started to revisit and modify the concept of time-space compression in the evolving social, cultural, and technological conditions of modernity. In 2003, Servaes and Wang revisit the time-space compression theory by saying, "Time and space are compressing but not eliminating."[18] The flows of capital and technology must eventually land in distinct places where people live their "local lives."[19] Moreover, it is common for human beings to long for a sense of belonging that is constructed by the sense of place and cultural identity.[20]

---

14. Benedict Anderson, *Imagined Communities: Reflections on the Origin and Spread of Nationalism* (London: Verso, 1983).
15. Harvey, "Conditions," 293.
16. David Harvey, *The Condition of Postmodernity: An Enquiry into the Origins of Cultural Change* (Cambridge, MA: Wiley-Blackwell, 1989).
17. Harvey, "The Condition," 147.
18. Jan Servaes and Georgette Wang, "Introduction," in *The New Communications Landscape* (New York: Routledge, 2003).
19. James H Mittelman, "How Does Globalization Really Work," in *Globalization: Critical Reflections* (Boulder, CO: Lynne Rienner, 1997).
20. Stuart Hall, "Cultural Identity and Diaspora," in *Contemporary Postcolonial Theory*, ed. Padmini Mongia (London: Hodder Arnold, 1996).

Though time-space compression envisions a world without temporal and spatial barriers, the sense of place and its connection to a local culture constitute the critical factors in the restructuring of the global communication industry.[21]

As one of the most common media, television is thought to have a significant role in the compression of time and space since "it constantly delivers distant events and concerns to people's homes and minds around the world."[22] In the aspect of technological character, television has undergone a series of revolutionary transformation, from broadcast to satellite and to current Internet streaming. As early as the 1970s, Williams foresaw that the rising satellite TV service would penetrate or circumvent national broadcast TV.[23] Similarly, in the early 2000s, the cord-cutting trend threatened to replace and cancel the relatively traditional cable video service and satellites.[24] In the new era of media communication, OTT platforms like Netflix and Apple TV+ further utilize Internet protocols to free the audiences from the then-national borders of broadcast TV, at the same time providing stronger Internet and more targeted programming than the satellites.[25] Streaming/OTT platforms emerge as the new site of encounter that connects temporality, spatiality, and globality. Despite the ongoing panic that accompanies technological transformations, OTT is never meant to kill television but to distribute television in an alternative way and to improve what we watch.[26] The current place of OTT platforms, like the former satellite beams and cable wire, can be seen in terms of a repeated pattern of technological evolution rather than a series of eliminations. The various ways of

---

21. Servaes and Wang, "Introduction," 5.
22. Servaes and Wang, 6.
23. Ramon Lobato, *Netflix Nations: The Geography of Digital Distribution* (New York: New York University Press, 2019); Raymond Williams, "Alternative Technology, Alternative Use," in *Television: Technology and Cultural Form* (New York: Routledge, 1974).
24. Amanda D. Lotz, *We Now Disrupt This Broadcast How Cable Transformed Television and the Internet Revolutionized It All* (Cambridge, MA: MIT Press, 2018).
25. Lotz, "We Now," 135.
26. Lotz, 5.

watching and distributing television across history will also affect the way we perceive time, space, movement, mobility, and cultural identities.

In terms of the transnational character, the concept of transnational television develops with respect to the advent of cable, satellite, and Internet streaming. For much of its history, broadcast television was closely tied to nationality and operated within national territories.[27] Starting from the 1970s, the emergence of satellite technology undermined the structure of national TV, and by the 1990s, the world media market dominated by national enterprise has been gradually eroded by transnational satellite television.[28] In the contemporary time, the OTT platforms redefine transnationalism as they construct a global media system that "takes television away from its national context, appealing to global audiences rather than national ones."[29] Simultaneous with the advent of transnational television, a cultural-linguistic market emerges after the dominance of American TV in the 1960s and 1970s.[30] The cultural-linguistic market, such as the increased demand for Latin American television, creates space for cultural products of diasporic and minority communities.[31] This emerging market in television-making is led by hybridization, as the method to blur the boundary between foreign and domestic.[32]

Like satellites have enabled long-distance market entry for Latin American television, OTTs have further expanded the transnational reach of Korean film and television—a regional cultural industry that has always been popular in regions of proximity has burst onto the global scene with great success in recent years. The above technological and transnational characteristics have enabled the broader production and distribution of K-dramas

---

27. Lobato, "Netflix," 62

28. Labato, 62; Servaes and Wang, "Introduction," 8.

29. Mareike Jenner, *Netflix and the Re-Invention of Television* (New York: Palgrave Macmillan, 2019).

30. John Sinclair, "Geolinguistic Region as Global Space: The Case of Latin America," in *The New Communications Landscape* (New York: Routledge, 2003).

31. Sinclair, "Geolinguistic Region," 8.

32. Servaes and Wang, "Introduction," 8; Sinclair, "Geolinguistic," 20.

and movies on global streaming platforms. Theresa Kang-Lowe, the executive producer of *Pachinko*, once expressed the difficulties of producing ethnic minority-themed dramas ten years ago and how she can finally execute today. As she says in the interview, "I have been waiting 20 years for the culture to catch up with my personal taste. The culture was not ready yet, unfortunately. I believe that we are here now. I do think these OTT platforms have equalized the playing field."[33]

TV Research as a Methodology to Understand Migration and Globalization

In recent years, the social-scientific turn to human migration and the rapid development of communication technology has called for interdisciplinary research that combines media studies with migration studies; for instance, digital migration studies examines the relationship between migration and digital connectivity.[34] In terms of television, the medium always has an intimate connectivity to our lived world, whether it is broadcasting news, educational shows, or TV dramas. As we often use the word *relatable* to capture a feeling after watching a show that speaks our mind, television can often better describe people's nuanced and intangible perception of the world they are situated in. Following TV's evolutionary trajectory as the starting point of the inquiry, this paper undertakes television research as the methodology to understand migration and globalization. It then closely analyzes *Pachinko* as a case study to describe the temporal, spatial, and linguistic qualities of the TV drama that are reflexive of migrant life.

---

33. "'Pachinko' Executive Producer Theresa Kang-Lowe on Korea's Hollywood Breakthrough: 'I've Been Waiting 20 Years for the Culture to Catch up with My Personal Taste,'" CNBC, March 2022, https://www.cnbc.com/video/2022/04/14/pachinko-executive-producer-theresa-kang-lowe-on-koreas-hollywood-breakthrough-ive-been-waiting-20-years-for-the-culture-to-catch-up-with-my-personal-taste.html.
34. Lee, "Temporal Dimension," 259

On the temporal dimension, TV series are typically twenty to sixty minutes in length and have at least two episodes, although some have several seasons that last for years. For *Pachinko*, instead of adopting the binge-watching mode, Apple TV+ chose to release the episodes on a weekly basis. The TV schedule simulates the common sensibilities of stoppage, waiting, and resuming that happen in a migrant's temporal perception. On the spatial dimension, the sensatory experience of sound, camera movements, camera distance and shot-by-shot transitions in *Pachinko* turn the abstract concepts of space and place into aural and pictorial sensibilities. Last but not least, the linguistic dimension. A show with an international ensemble cast and considerable loads of dialogues, translations, and subtitles encourages audiences to follow the linguistic and cultural conventions of the local, distinct, or cultural-specific places. In order to explain the three dimensions of migration in *Pachinko*, this paper focuses on two types of storytelling: the first is visual storytelling that conveys time-and-space perception by implementing television form and style; the second is dialogue storytelling that conveys linguistic multiculturalism through verbal communication. The temporal and spatial dimensions not only look at methods in film and television studies but also theories of human migration and geography. The third dimension, which moves on from visual storytelling to dialogue storytelling, delves into the metaphorical and symbolic representation of human languages during migration and globalization.

Spatial Dimension: Space, Place, Placelessness, and Displacement

The concept of spatiality in *Pachinko*'s migration trajectories is not only about the geometric abstraction of space but also the direct experience and relationship associated with places. Edward Relph suggests that the meaning of space, particularly the lived space, comes from the "existential and perpetual places of immediate experience."[35] Heidegger also argues that spaces receive their being from places, not the space; and dwelling as the essential property of human existence is constituted by humans' relationships to places.[36] Based on Relph's theories of space and place, I draw the correlation that in a migrant life, the meanings of place often involve rootedness, the home place, and placelessness.[37] Rootedness in a place is associated with the communal and personal experience of knowing and being known in the place, and having a root means long-lasting caring for the place. For example, Yeongdo in Korea is the root of the family because it is their homeland and locus of heritage. The home place is "a particular setting in which we are attached" and a point of departure "from which we orient ourselves in the larger world."[38] Osaka is not the root of the family in *Pachinko*, but they develop a new connection to the place as they make their homes there. Placelessness, on the contrary, describes "the monotonous, shallow and placeless flatscape that lacks intentional depth," particularly coming from "the undesirable, inauthentic aspect of the modern age."[39] Though geographical uniformity is not a new phenomenon, the scale of placelessness is expanding partly because of media communication. The way media affects placelessness is similar to time-space compression. As Relph argues, "Media communication includes television, radio, journals and newspaper and other media that have reduced the need for face to face contact and freed communities from their localities. They tend to report problems as general and widespread rather than local and specific."[40] In *Pachinko*, there are representations of rootedness and the home

---

35. Edward Relph, *Place and Placelessness* (London: Pion, 1976).
36. Martin Heidegger, "Building Dwelling Thinking," in *Poetry, Language, Thought* (London: Harper & Row, 1971), 143–61.
37. Relph, *Place and Placelessness*, 37.
38. Relph, 39.
39. Relph, 79–80.
40. Relph, 90.

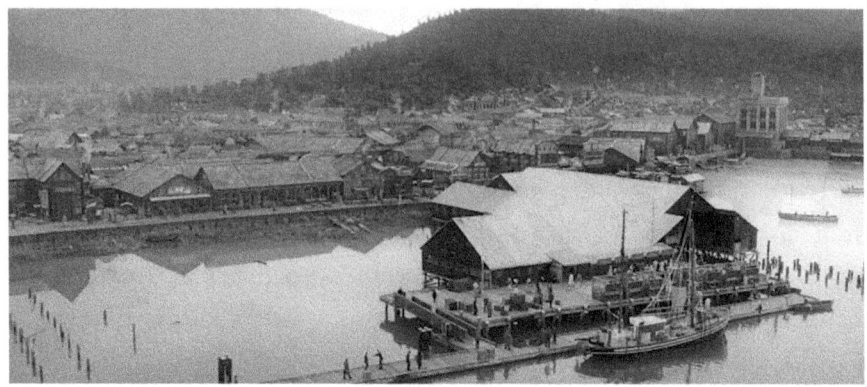

Figure 1: Yeongdo, Busan, Korea, in *Pachinko*.
Source: Apple TV+

place, such as the Busan fishing village by the sea where the main character was born in 1910 and the muddy immigrant neighbourhood in Osaka where the character relocates in 1931. There are also placeless scenes in the show such as skyscrapers, offices, and banquet halls in 1980s Tokyo and New York City.

Apart from place and placelessness, another significant experience with space and place is the sense of displacement. As mentioned by Robert Tally, displacement "underscores the critical importance of spatial relations in our attempts to interpret and change the world."[41] When time-space compression generates new geographies over time, people are constantly landing in new places, losing old places, and being displaced. The sense of displacement fills the entire series in *Pachinko*, whether it is manifested in the characters' lives or reflected in the audience's viewing experience. *Pachinko* tracks the characters' life journeys across generations and locations, mapping the spatial changes through title pages, camera movements, and noncontinuity editing. The camera pans through four

---

41. Robert T. Tally, *Spatiality* (London and New York: Routledge, 2013).

places, Busan, Osaka, Tokyo, and New York City, in wide soaring aerial shots as the language of the epic, implementing special effects such as superimposition to transit from one locality to another. Title pages in three languages—English, Japanese, and Korean—overlay the varying magnificent landscapes. As the audiences receive the televisual techniques, they are immersed in the same sense of displacement as the characters and in constant questioning of where they are in this show. When television narratives help us transcend spatial and temporal barriers, they have the power to "revolutionize the objective qualities of space and time that we are forced to alter."[42] Fitting multiple plotlines in one metanarrative hints at the problem of not having enough time to thoroughly experience a single event at a distinct place, at the same time mirroring the disorienting experience of a migrant.

Temporal Dimension: Temporality on Screen and Temporality in Migration

History, narration, and stories all take place in time. Film and television use time to shape our understanding of narrative action. In *Film Art*, Bordwell uses temporal order, duration, and frequency to define the variants of time in a film that are subject to manipulation.[43] In a similar vein, Mary Ann Doane disentangles filmic time based on its subjective perception, calling it the temporality of apparatus, temporality of diegesis, and temporality of reception.[44] In television terminology, Jason Mittell distinguishes three types of time in all narrative works. Story time refers

---

42. Harvey, "Condition."
43. David Bordwell, Kristin Thompson, and Jeff Smith, *Film Art: An Introduction* (New York: Mcgraw-Hill, 2015).
44. Mary Ann Doane, *The Emergence of Cinematic Time: Modernity, Contingency, the Archive* (Cambridge, MA: Harvard University Press, 2002).

to the time frame of the diegesis; the time passes through the story world in the manner of the real world. Discourse time is the temporal structure and duration of the story as told within a given narrative. Narration time is the total duration of telling and receiving the story, such as a two-hour film, a ten-episode series, or a ten-chapter novel.[45] The theories of time in cinema and television share a commonality in that the time experienced in the real world is often different from the time perceived in the narrative world. Narrative works, including literature, films, and television series, have the long-standing practices of manipulating temporality by skipping, repeating, or prolonging story events. It has become a creative strategy to reimagine time from its objective and quantitative dimensions. In other words, film and television are also the most convenient media in compressing time with their storytelling techniques and visual and aural languages.

Combining the previous conceptualization of time by Bordwell, Mittell, and Doane, I describe two types of time in *Pachinko*. First, the time of the story world. It refers to the objective time of eighty years that Sunja's family tree has grown, spanning the lifetimes of three generations and the main character Sunja's transformations from a little girl to a grandmother. Second, the audience's time. It is the time experienced by the audience as they watch the entire eight episodes of the show. On the screen, the uneventful time spent by the characters—such as cooking, sleeping, commuting, and working—is skipped, hence the decades of family history are trimmed to forty-five minutes each week in a two-month streaming period. Audiences can also perceive the time of particular events as much longer or shorter than they would through various techniques such as fast and slow motions, repetition, and elongation. Moreover, the audience's time is deliberately confused by the plot as *Pachinko*'s narrative is in nonlinear story order. The audiences are constantly "time-travelling" in the show—living

---

45. Mittell, "Complex TV."

in the 1910s with Sunja in the past minute and situating themselves with Solomon in 1980s New York City in the next—all by a subtle switch of shots.

To bridge the temporal connections, I suggest that the temporality of media is reflexive of the temporality of migration, as they are both subject to human experience. The temporal dimension of migration concerns time and temporality differently—time refers to its quantitative, objective nature and temporality refers to the subjective experience of time. Time is usually manifested in "everyday work, leisure and media time, or the length of time associated with a visa," but temporality shows how time passes differently for migrants when they are waiting and being stuck, without the legal status to work and live properly in the country.[46] On the other hand, in the world of screen media, there are many techniques and visual languages that intentionally make events feel shorter or longer than the events normally last. Common stylistic choices are fast and slow motion, skipping and repetition, pauses and interruption, etc. In episode 7 of *Pachinko*, the character Hansu is thrown into a cataclysmic disaster, the 1923 Great Kanto Earthquake. The shock only lasts momentarily but the episode uses slow-motion camera movement, point-of-view shots, and subjective sound to emphasize the long-lasting effects of the earthquake—a quick but fatal shock that took thousands of lives away and perpetually changed everything about the people and the land. For about twenty minutes of the episode, we see Hansu stumbling on the street, searching for his friends, dodging fallen rocks, waking up from a coma, and witnessing more people die around him. The rest of the episode feels like a walk that never ends, with massive fears from the subjective perception of time and space created by images and sound. As the temporality of migration is subject to how migrants internalize wait time and stoppage, the temporality of media is subject to how the writers and artists depict the events and how the audiences experience them on screen.

---

46. Lee, "Temporal Dimension," 260.

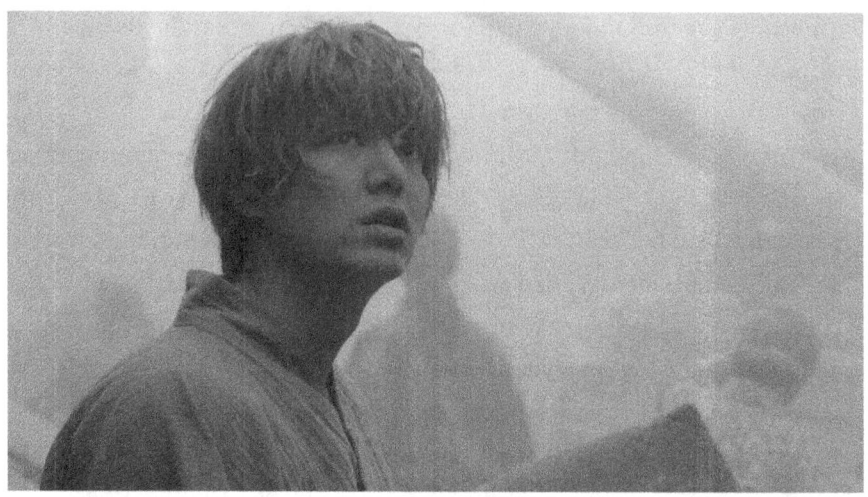

Figure 2: Hansu, performed by Lee Min-ho, after the 1023 Great Kanto
Earthquake, episode 7.
Source: Apple TV+

Linguistic Dimension: Language, Translation and Multilingualism

Language is a major barrier in migration because the incapability to speak
cuts off the possibility to assimilate and integrate. When people migrate, they
are compelled to pick up new languages in order to start new lives; meanwhile, they also try to remember old languages as the most direct connection
to their heritage. Eventually, people utilize multilingualism and translation
to facilitate inconvenient situations due to language barriers. This section
of the essay focuses on the language problem in migration—particularly
the question of whether it provokes personal trauma or fashions cultural
identities—in an environment dominated by discrimination and racism.

In the Christian tradition, speaking multiple languages bears negative
connotations that stand for catastrophe, punishment, and curses. In the
Bible story of the Tower of Babel, all humanity lives in one single place and

speaks one common language.[47] Human beings communicate and collaborate very well in a single language; therefore, they began the ambitious project of building a massive tower that leads to heaven. God is offended by the power of humanity. He puts an end to the project by deliberately confusing the common human language, causing them to speak many different languages so they would not understand each other. Losing the tool to properly communicate, humans abandon the tower, return to the earth, and speak different languages thereafter. The Tower of Babel is the mythic origin of the human population being scattered in different countries and disconnected by language barriers.[48] It is God's way to punish humanity for usurping his omnipotent power.

Lydia Liu further discusses the problem of language in cross-cultural literary studies in her work[1] *Translingual Practice.* She particularly refers to the story of Babel as a symbol of the chaos of human communication. In the countless versions that have been circulated, the Babel story not only suggests "the impossibility of translating among the irreducible multiplicity of tongues but also projects a desire for completion and for original Logos."[49] However, the story of Babel is contradictory in nature because we read about this narrative through translation, yet it continuously emphasizes the failure of collaborating across different languages. Aside from Babel, Liu discusses the translator's active negotiation between the "source language" and the "target language." She urges us to consider "in whose term, for which linguistic contingency and in the name of what knowledge someone is translating between cultures."[50] This question is also applicable to cinema and television since literary texts became more convergent. For instance,

---

47. Genesis 11:1–9.

48. David J. Gunkel, "Machine Translation Mediating Linguistic Difference in the Era of Globalization," in *The Routledge Handbook of Digital Media and Globalization* (New York: Routledge, 2021), 198–205.

49. Lydia He Liu, *Translingual Practice Literature, National Culture and Translated Modernity—China, 1900–1937* (Stanford, CA: Stanford University Press, 1999).

50. Liu, *Translingual Practice Literature*, 2.

the source language in *Pachinko* is Korean and the target languages are the over forty languages of translated subtitles provided by Apple TV+. The options appear to be abundant. However, the positions of languages become more ambiguous in dubbing: only Korean dialogues are dubbed and the Japanese and English dialogues remain original; mainly Anglophone and European languages are provided in dubbing, such as French, Spanish, Portuguese, and Italian. The dubbing options demonstrate that first, the acts of aural translation are still in Western terms, possibly due to the origin of the platform; second, the Japanese language holds a strange position, which is neither original nor targeted, in the translation of this show. Thus, the strange position of the Japanese language coincides with the characters' underlying psychology of not having a sense of belonging to Japan.

The complex consequences of speaking multiple languages as one's identity are insinuated in *Pachinko* multiple times—not just literally but also metaphorically. The following analysis examines how speaking multiple languages can enable or perhaps burden someone in navigating their migration trajectories. I analyze a scene in episode 8 as an example, when Noa translates for his mother in a critical situation. Noa is the child of a Zainichi woman in Osaka. At a very young age, he has to be the translator of the family because Sunja, his mother, does not speak Japanese. In the scene, Sunja's husband is arrested for a mysterious cause, and Sunja has to investigate and brings Noa with her as a translator. Sunja is introduced to Mr. Hasegawa, who has information about her husband's arrest. When Noa and Sunja enter the room, Mr. Hasegawa and his daughter are sitting in a dark corner. Noa walks into the room and speaks to Sunja in Korean, "He said come in and sit." Then, Mr. Hasegawa speaks while Noa translates. In the beginning, the conversation is still calm, polite, and mundane. Here, Noa's role is merely a translation machine, keeping his distance from the content of the conversation. As the conversation escalates, Sunja is stunned by her husband's secret life as a communist. She then starts to fluster, scream,

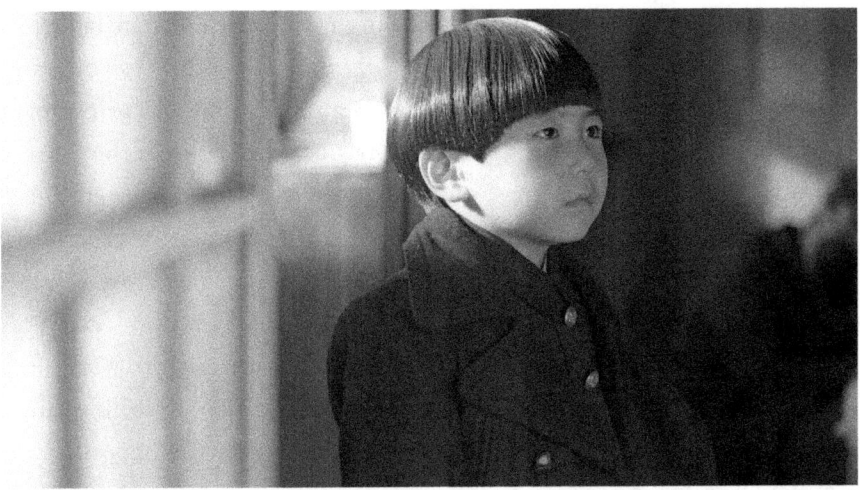

Figure 3: Noa, performed by Park Jae-joon, appears in episode 8.
Source: Apple TV+

and weep. Without much capacity to comprehend the actual situation, Noa can only understand that his father is in danger through his mother's emotional expressions. He then stands up and yells at Mr. Hasegawa, "You are lying!" Then, the scene ends with a group of police breaking into the room.

The scene links translation with catastrophe and trauma in a number of ways. First, when Noa translates, he is at the center of the "chaos of human communication" because he is the only person who can perform as a common-language mediator. Metaphorically, he becomes the lone man in the Tower of Babel.[51] As the phrases get harder, he faces the impossibility of translating among the "irreducible multiplicity of tongues."[52] Second, he feels the sense of displacement of a second-generation immigrant who relocates and is disoriented between cultures. Noa is only a

---

51. Liu, 11.
52. Liu, 11.

seven-year-old child who has limited linguistic skills. For instance, he could barely understand the word *communist*. In the process of translating, he is gradually exposed to the facts about his father's arrest, even before his mother has access to the information. He stands at the front of a family catastrophe when he should be the one protected from it. Third, we see the illiteracy of the audience. Apple TV+ intends not to sync the translation of the Japanese dialogues when the Hasegawas are talking. As non-Japanese speakers, we are as ignorant as Sunja and can only depend on Noa's translation. We thereby further sympathize with Noa's situation of being exposed to trauma firsthand. To Noa, multilingualism is a situation and a result of his cultural identity. As the offspring of a Zainichi family, he has to master both languages to support the family to survive in a discriminating society. The later story alludes to the fact that Noa develops post-traumatic reactions after he grows up and eventually commits suicide.

As the show writer Soo Hugh expresses in an interview, "Language defines a huge part of who we are and how history is unpacked. Switching language tells so much about a person, how the character speaks that language makes a character complete, or how the language is being lost."[53] Solomon, who appears to be a grown-up version of Noa, also masters most languages in this show. Language is a tool for Solomon, as he can speak the Korean dialect to convince the elderly neighbors to sell the land; he can use native Japanese to communicate with his Tokyo coworkers; he can use English fluently to negotiate deals with his white bosses. Multilingualism reflects the character's talents, capacities, and status—as Solomon is often considered "the one who made it" in the family by attending Yale University and evolving into the elite. However, multilingualism is also accompanied by insecurities in the constant pursuit of external validation. Being called a pachinko owner's son throughout his formative years, Solomon bears

---

53. "'Joy Luck Club' to 'Pachinko,'"

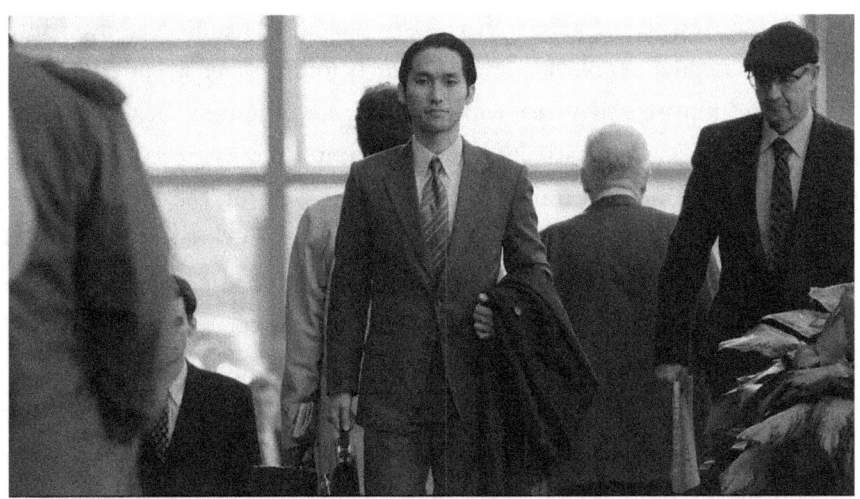

Figure 4: Solomon, performed by Jin Ha, is walking in a building at Wall
Street in New York City, episode 1.
Source: Apple TV+

generational expectations and strives to shed the shame of a "Zanichi's Offspring." He is benefited—and also burdened—by his exceptional ability to
assimilate. As cultural identities get more complicated, contradictory, and
relentless, one cannot simply resolve the identity crisis by assimilation and by
denial of the past. Seamlessly transitioning between English, Korean, and Japanese, Solomon's language skill is the symptom of adjusting and assimilating to
a global metropolitan culture in a shrinking world. This interpretation finally
leads to the conclusion of this paper, calling for an understanding of navigating
between K-drama and Hollywood and between barrier and assimilation.

Conclusion

As Hale puts it in a *New York Times* review, "Hollywoodization, voluntary
or not, is the operative word when it comes to both '*Joy Luck Club*' and

141

'*Pachinko.*' And to the extent that glossy melodrama pulls audiences into a story that puts people we haven't seen before onscreen, and treats the hatred and injustice they face with some degree of honesty, it's not a dirty word."[54] The review is partially right. It indicates that criticism is less concerned about American imperialism in the new era of TV, because many transnational programs today are learning to adapt to each other instead of the one-way copying of the American-centric model. *Pachinko*, with the ambition of bringing East Asian history to Hollywood, is a show that reflects on the collective endeavor in multicultural and multilingual storytelling. It maintains the "soap-operatic" appeal of a K-drama and at the same time adapts to the narrative conventions and audiovisual language in American TV.[55] Thus, *Pachinko* is rather hybridized than Hollywoodized by being able to balance the elements between the global and the local.

Film and television can give us a world by creating unique viewing experiences with its narrative forms and styles. However, condensing eighty years of history into eight episodes leaves many stories unspoken. A year after the premiere, season 2 is in the making, but the audiences' attention is harder to sustain in the streaming era. With an abundance of transnational programs and a short turnover time, the global digital platform epitomize the time-space compression we are currently experiencing. We as viewers and perceivers are given enough choices to time-travel with streaming series from here and there, now and then, but we are given less time and space to react to the whole picture of history. On the other hand, the disorienting viewing experience, whether it is because of the frequent relocation or switch of languages, mirrors the uprooted lives in *Pachinko*. The current time-space compression constantly creates new geographies and imaginations; meanwhile, it evokes a more complex combination of feelings such as excitement, confusion, stress, and anxiety.

---

54. Hale, " '*Pachinko*' Review."
55. Hale.

Pachinko makes a significant move by maintaining the original languages. For a long time, audiences have borne the out-of-context English dialogues in non-Anglophone films and TV. A counter-example is Netflix's *Marco Polo* (2014). Despite the fact that the TV show is in foreign settings and the actors are able to speak the local languages, the story is still told in English dialogues. The use of language implies that the production is only made for English audiences, which makes *Marco Polo* a "cross-cultural clunker."[56] The inevitable trend of transnational TV may change the orientation of original-language shows. Besides, *Pachinko*'s multilingualism is accredited to its much more precise translation and subtitling service. The subtitles in *Pachinko* not only translate but also give the personal, cultural, and historical contexts of the language being spoken. From the audience's perspective, listening to the original languages provide more backstories and depth to the characters. In the Tower of Babel, the dream of a common language was taken away by God. With the various options in subtitles and dubbing provided by streaming platforms, humanity seems to have the ability to collaborate on the Babel project again. Aside from streaming without national constraints, multilingualism and subtitle viewing have opened up more space for transnational cultural production.

Acknowledgments

The author wishes to thank professor Ying Zhu and professor Tze-lan Sang for their support of this paper, starting from its draft that was presented at the Associations for Asian Studies annual meeting in 2023 through the final completion of the manuscript. The author also wishes to extend her gratitude to the anonymous reviewers for their invaluable feedback and her special thanks to Alvin Luong for his editing support.

---

56. Bethany Allen-Ebrahimian, "Netflix's 'Marco Polo' Is Cross-Cultural Clunker," Foreign Policy, December 2014, https://foreignpolicy.com/2014/12/24/netflixs-marco-polo-is-cross-cultural-clunker/.

Drama Reviews

Squid Game

The Hall of Screens in the Age of Platform Cosmopolitanism

MEI MINGXUE NAN

Abstract

The nine-episode Korean-language series *Squid Game* became a global sensation immediately upon its premiere on Netflix in September 2021. The show's popularity and critical acclaim in the anglophone world had been unprecedented for a non-English series. This review provides a symptomatic reading of *Squid Game*'s global success and a short analysis of its visual appeal. It also explores the tension between *Squid Game*'s smooth and flat aesthetics that enables the show to travel and the culturally specific contexts it references. Some argue that the aesthetics lead to an ahistorical and superficial cultural understanding that overlooks the complexities of Korea's history and US imperialism. Others argue that they challenge cultural hierarchies and democratize interactions. This reflects a broader challenge of balancing global accessibility with cultural specificity faced by East Asian serial dramas in the era of global streaming services. This review concludes by highlighting the role of platform cosmopolitanism in bridging cultural and linguistic barriers in meaningful ways.

Keywords: *Squid Game,* Global Streaming Services, K-drama, Surveillance and Spectatorship, Platform Cosmopolitanism

https://doi.org/10.3998/gs.4156

"We knew we wanted this show to travel," says Minyoung Kim, Netflix's vice president of content for the Asia-Pacific region (excluding India).[1] Invited by *The Hollywood Reporter* to explain *Squid Game*'s global appeal, Kim comments that the show is "perfect evidence that our international strategy has been right." The nine-episode Korean-language series became a global sensation immediately upon its premiere on Netflix in September 2021. As of this review's publication, it is the platform's most popular show of all time based on hours viewed in the first twenty-eight days of release.[2] It is also the first non-English TV series to be nominated for and win Primetime Emmy Awards. Such popularity and critical acclaim in the anglophone world had been unprecedented for a non-English series.

Written and directed by filmmaker Hwang Dong-hyuk, the show revolves around a survival game where 456 destitute social outcasts compete in six children's games for a massive 45.6 billion won (over $30 million USD) cash prize. In the first episode, the audience follows the backstory of the protagonist Seong Gi-hun (played by Lee Jung-jae), a debt-ridden gambling addict who had lost custody of his daughter. Desperate to redeem himself and cover his mother's medical expenses, Gi-hun finds himself lured into a lethal series of seemingly innocent children's games. The first game resembles the classic "Red Light, Green Light," where the contestants face an enormous motion-sensing animatronic doll that announces each violation and subsequent death with a disconcertingly childlike voice (Figure 1).

The distorted version of "Red Light, Green Light" went viral on social media, with memes flooding in under #SquidGame. The show quickly rose to fame across the world. Netflix, individual content creators, gaming

---

1. Patrick Brzeski, "'Squid Game': Netflix's Top Exec in Asia Explains the Show's Huge Global Appeal," *The Hollywood Reporter*, October 11, 2021, https://www.hollywoodreporter.com/tv/tv-news/squid-game-secret-to-global-success-1235030008/.

2. "Most Popular TV (Non-English)" Netflix, accessed April 5, 2023, https://top10.netflix.com/tv-non-english?week=2023-02-12.

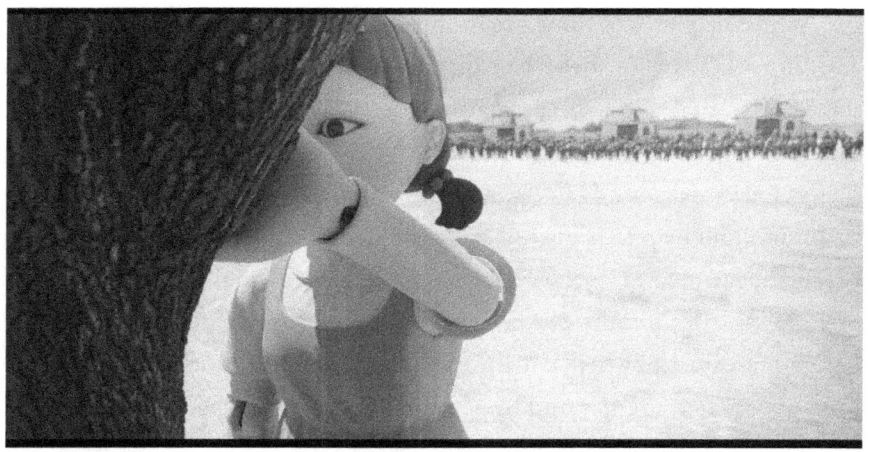

Figure 1: The iconic doll
Source: Screenshot from *Squid Game* on Netflix

companies, and local businesses have been swift to capitalize on its success, from themed cafés in Paris and Chengdu to replicas of the massive robot doll in Manila and Sydney, from a Minecraft live-stream attracting 200 participants to play together to hourlong immersive adventures provided by Immersive Gamebox in the US, UK, Germany, and United Arab Emirates.

Squid Game's global success may have been unexpected but certainly welcomed. During her interview, Kim attributes the success to five key elements:

1. The popularity of the survival game genre among global audiences
2. The cultural authenticity and relative simplicity of the games
3. The art, especially the mise-en-scène and the music, enabled by big budgets
4. The memorable, meme-able moments to drive conversations
5. The overarching message on the universality of social injustice, conveyed in an entertaining way

While debates continue regarding the claims of cultural authenticity and universality against the backdrop of uneven globalization, there is no doubt that *Squid Game* is cinematically well-crafted. The meticulously laid out mise-en-scène featuring surreal *Alice in Wonderland*–style visuals, the perky music and sound effects, and the mischievous graphic match cuts to add humor all help to achieve a stark contrast with the bloody and ruthless reality for each player. *Squid Game* is a darkly playful doubling of our neoliberal capitalist reality: the powerful and rich design and gain from an inherently unfair system, where cut-throat competitions are framed as games that everyone is "free" to play and to potentially win under the watching eyes of surveillance and spectatorship.

Hall of Screens: The Omnipresent Cameras and the Oscillation of Perspectives

It is interesting how *Squid Game* constantly reminds its viewers of the intricate layers between the observer and subject under the ubiquity of cameras and screens, inviting Netflix's audience to reflect upon their own positions in the act of looking. The show provides a visceral experience of surveillance and spectatorship featuring both human and nonhuman observers. By cycling through different perspectives, *Squid Game* begs the question of with whom do we identify in each situation. A player? A worker? An on-screen or off-screen spectator? An inanimate object? As a result, the show creates a multilayered viewing experience, where the audience's identification oscillates through multiple perspectives from looking through different lenses.

There are several levels of human observers, from the contestants to the workers to the front man to the VIPs, further complicated by the audience's external view from the computer screen. Figure 2 is a collage showing how viewers are invited into the story world. Both shots employ the composition technique of a frame within a frame, which is often used to highlight the subject being observed. In the case of *Squid Game*, it has an additional effect

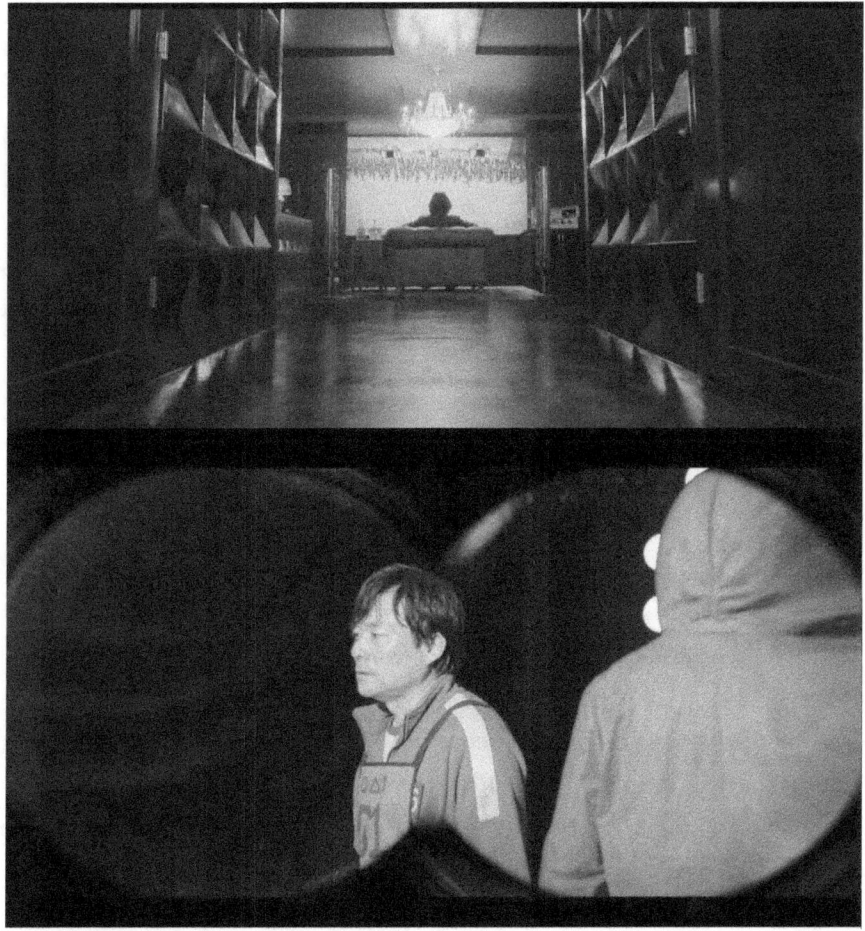

Figure 2: Frame within a frame
Source: Screenshot from *Squid Game* on Netflix

of drawing attention to the act of observation itself. In the top image, the viewer is positioned down the hall behind the front man as he monitors the contestants. The shot is a reminder that the very act of watching, regardless of sitting in a surveillant's lounge or at home on the couch, entails a certain degree of voyeurism. The bottom image from episode 7 conveys this

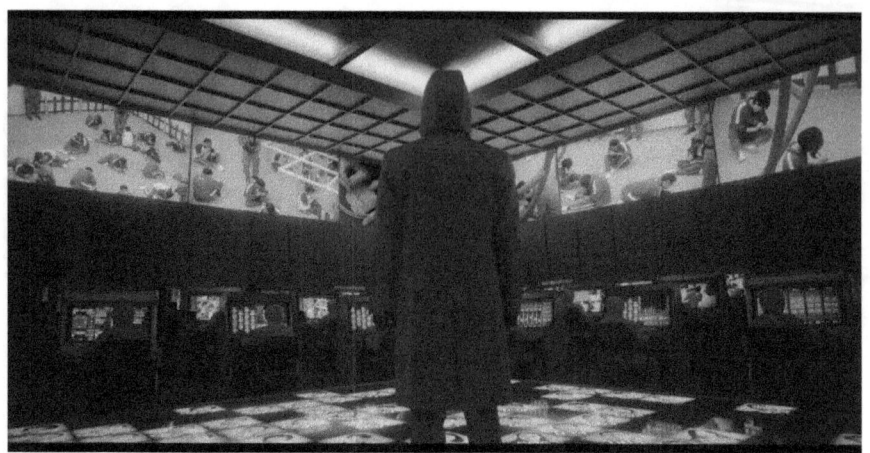

Figure 3: The control room as a hall of screens
Source: Screenshot from *Squid Game* on Netflix

message even more strongly, as the magnified point of view (POV) puts the audience directly in the position of a VIP looking through binoculars.

While the most explicit reference to voyeurism and spectatorship occurs within the VIP lounge, the recurring scenes of the control room highlight the theme of digital mass surveillance. Figure 3 shows a comprehensive view of the room, packed with screens from top to bottom with the photo grid floor, manned computer stations, and the panoptic surveillance monitors on the walls. The viewer watches the front man oversee the workers who monitor the contestants, forming a chain of looking. All these human observers, including the audience, are trapped in an infinite hall of screens while surrounded by myriad digital devices. In this sense, the show makes an ambivalent comment on the power dynamics of spectatorship in the age of neoliberalism as well as the totalizing power of media technology that shapes how we see and are being seen.

Compared to shots from human perspectives, the POV shots from the viewpoint of automatic machines—which I call the *posthuman shots*—are even more intriguing. The posthuman shots reverse the typical roles of who is viewing whom in a machine-human relationship, thus evoking an uncanny feeling. Some of these shots are closely spaced during the last twenty minutes of the first episode, which is a sequence of parallel editing between the players and the front man. One of the most memorable parallels is between two shot/reverse-shot sequences one between Gi-hun and the camera at the self-service registration kiosk; the other between the front man and the door camera with a facial recognition security system (Figure 4). Conventionally, a sequence of over-the-shoulder shots/reverse shots are often used to create an emotional connection between two characters who are engaging in a conversation. As the camera alternates from one character to the other, the audience is supposed to empathize with both. In contradistinction, the shot/reverse-shot sequences here do not occur between two human or anthropomorphic characters but between people and impersonal machines. The POV shot of an automatic machine looking at a person provides a unique, nonanthropocentric perspective. The effect is a kind of posthuman intimacy,

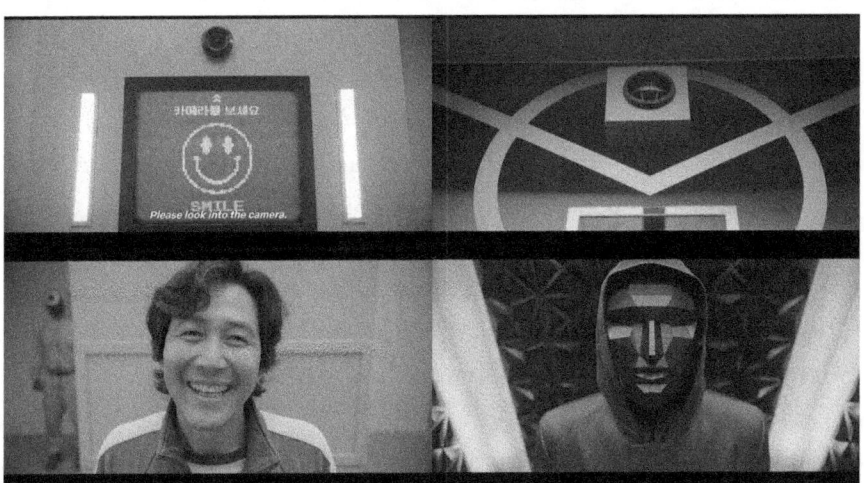

Figure 4: The posthuman intimacy between automatic machines and humans
Source: Screenshot from *Squid Game* on Netflix

153

where our everyday interactions with the cameras of phones, laptops, security systems, and inspection kiosks are reconfigured into machine-human conversations. However, the show also uses posthuman shots to demonstrate the nonhuman observers' disinterest in human lives. From the eyes of the motion-sensing doll to the high angle perspective of the automatic gun turrets, human beings are reduced to mere data points (Figure 5).

It follows that with the omnipresent cameras and the oscillation of perspectives, *Squid Game* comments on the issue of looking from at least three aspects:

1. The power dynamic of voyeurism against the backdrop of global capitalism and the transnational capitalist class
2. The omnipresence of digital surveillance
3. The question of spectatorship in the age of global streaming services

While the hall of screens can be a metaphor for panoptic surveillance, it can also function as a metaphor for the thumbnail grid on video streaming platforms. In the latter interpretation, images from every corner in the world are brought together for consumption. The audience sitting in the hall of screens experiences what I call a *feeling of platform cosmopolitanism*, where a viewer digitally travels around the world unencumbered by cultural and linguistic barriers, experiencing curated tidbits of a life elsewhere.

The Platform Is the Message? East Asian Drama Series in the Age of Platform Cosmopolitanism

With the rise of global streaming services such as Netflix, travel has become something quite different. Gone are the times of *Around the World in Eighty Days* à la Jules Verne or *Around the World in Eighty Books* à la David Damrosch. In the age of platform cosmopolitanism, a Netflix user can hop around the world in eighty seconds, skipping from *Squid Game*

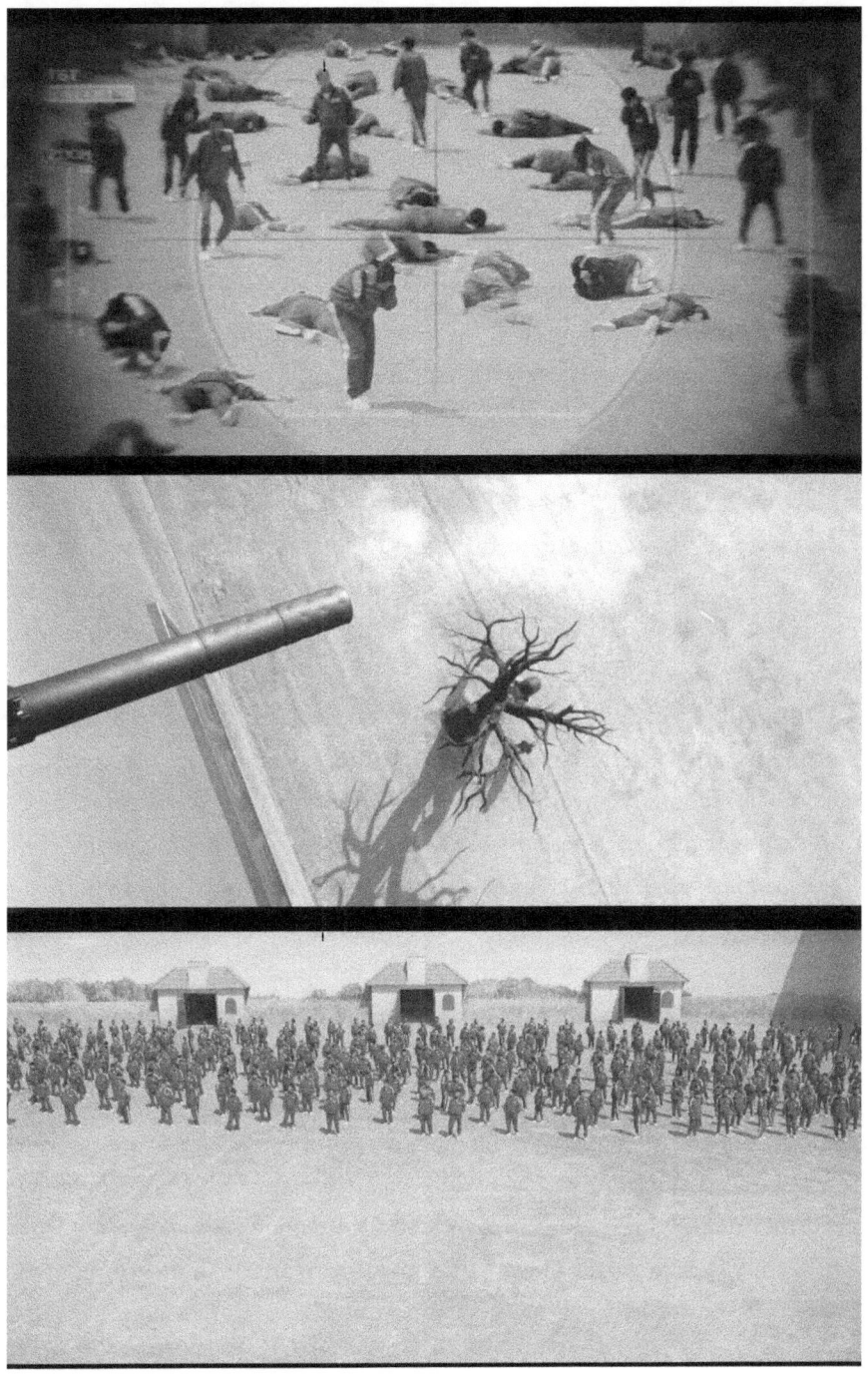

Figure 5: Posthuman shots seeing the contestants via the doll and the gun terrets
Source: Screenshot from *Squid Game* on Netflix

(contemporary South Korea) to *Stranger Things* (1980s North America) to *The Woman King* (1800s West Africa). The technology of digital streaming platforms has fundamentally changed the everyday experience of time and space by collapsing them into a feeling of right now, right here. To appropriate William Blake's famous lines in a cynical manner, browsing Netflix on a cell phone screen truly enables one to hold infinity in their hands and experience eternity in an hour. Marked by accessibility, flexibility, and seemingly endless choices, digital streaming platforms provide the audience with content that is meant to be consumed as a distraction, whether it be for entertainment or socialization purposes. Circulation value (cultural capital) thus outweighs representational value (cultural understanding), calling for content that follows the aesthetics of the smooth and flat, an aesthetics that *Squid Game* exemplifies.

Smoothness means no delay. Netflix endeavors to provide the audience with a smooth viewing experience (i.e., automatically skipping credits to encourage binging behavior, releasing non-English titles with multilingual options for anglophone viewers to watch in their preferred language, etc.). Flatness means bringing things of different distance onto the same plane (my term choice here is inspired by Murakami Takashi's viral concept "superflat"). It implies the superficiality of our postmodern consumer culture, aided by media technology collapsing time and space. Both the smooth and the flat indicate an erosion of cultural linguistic boundaries with the streaming platforms' global reach, achieved by a strategy of simplification. An example of this strategy would be how *Squid Game* made sure the rules of the Korean children's games featured are simple enough for people who are not familiar with those games to follow and enjoy.

Game aesthetics and video game logic, with their ability to transcend the boundaries of national markets and cultures, seem to be another common element shared by many films and drama series in this newest revival of the Korean wave. Examples include survival horrors featuring zombies and other monsters from *Train to Busan* (2016) to *Kingdom* (2019) to *Sweet Home* (2020), and, after the global success of *Parasite* (2019), the marrying of the survival game genre to social commentary in *Squid Game*, *Hellbound* (2021), and *All of Us Are Dead* (2022). Even for *Parasite*, Bon Joon-ho comments on how the design of the Park family's house—a dungeon for the Kim family to roam around and explore—resembles a video game.[3] These films and drama series are also examples of the blurring of big screen (cinema) and small screen (television) as a consequence of the rise of global streaming platforms. Digital platforms bridge the gap between cinema and TV, which leads to film aesthetics and techniques being transported into the production of popular drama series (e.g., the zombie genre from *Train to Busan* to *Kingdom*, the violence and dark humor from *Parasite* to *Squid Game*). This process is particularly visible in Netflix's strategy to harness the critically acclaimed, international image of South Korean cinema to breathe new life into K-dramas, drastically changing the latter's image from being full of romantic clichés to offering sharp social critiques.[4]

The smooth, the flat, and the blurring of boundaries are some defining features of platform cosmopolitanism, which goes hand in hand with neoliberal imaginations of global connectivity, free-flowing capital, and a cosmopolitan identity that hinges upon the feeling of relatability. Bon's famous comment that "we all live in the same country, it's called capitalism" encapsulates the media situation of platform cosmopolitanism we live in and the type of content it calls for.[5] A non-English title's global success (*Squid Game* as a prime example) depends on how well it reduces cultural barriers to appeal to an international audience. The international audience then

---

3. Todd Martens, "From '1917' to Yes, 'Parasite,' Video Games Are Even Influencing Prestige Movies," *Herald and News*, February 13, 2020, https://www.heraldandnews.com/from-1917-to-yes-parasite-video-games-are-even-influencing-prestige-movies/article_0710ef90–4e96–5530-a4f6–87cde4445c7f.html.
4. Yaeri Kim, "'Funny, Political and Bone-Crunchingly Violent': Squid Game and the Unintended Nation Branding of South Korea" (presentation, Association for Asian Studies Annual Conference, Boston, MA, March 17, 2023).
5. Kate Hagen, "The Black List Interview: Bong Joon-ho on Parasite," *Black List Blog*, October 11, 2019, https://blog.blcklst.com/the-black-list-interview-bong-joon-ho-on-parasite-5fd0cb0baa12.

receives a sense of relatability, a cosmopolitan identity consuming foreign media objects and navigating foreign cultures by simply browsing Netflix.

The power of the smooth and flat is evidenced by how anglophone journalists and content creators rave about *Squid Game*'s relatability. However, in a special forum on *Squid Game*, Raymond Kyooyung Ra makes a pointed argument against the idea of relatability. Recounting the show's references to locally specific contexts such as the 2009 SsangYong Motor strike, which are conveniently overlooked by the Hollywood press, Ra argues that "consuming *Squid Game* with little to no background knowledge of Korea as well as the United States' imperialist influences there" makes the American spectators' viewing experience "but a pleasurable act of appropriating culture capital."[6] Other forum contributors also express concerns over how Korea-specific contexts might get lost in translation. For example, David C. Oh points out that the Korean name of the first game does not translate to "Red Light, Green Light" but is rather based on the phrase of "the Rose of Sharon has bloomed." With the Rose of Sharon being South Korea's national flower, the name implies the Korean nation as a site of hopelessness and death.[7] For the sake of smooth circulation, something gets lost in translation and cultural understandings become flat. At the end of his article, Ra questions *Squid Game*'s complicity in aiding US viewers' ahistorical consumption, displacing South Korea's regional problems under US imperialism to universal class struggle, which implicitly challenges Minyoung Kim's claim to the show's cultural authenticity, universality, and profundity.

While these are valid concerns, it seems to go back to the long-standing question of text versus context. It sees the production and consumption of the show as a process of encoding/decoding, where area-specific knowledge is

6. Raymond Kyooyung Ra, "At the Center of Its World, the U.S. Empire Forgets itself: *Squid Game* and the Hollywood Press' Melodramatic Gaze," *Communication, Culture and Critique* 15, no. 4 (2022): 546–48, https://doi.org/10.1093/ccc/tcac040.

7. David C. Oh "The Politics of Representation in *Squid Game* and the Promise and Peril of Its Transnational Reception," *Communication, Culture and Critique* 15, no. 4 (2022): 531–33, https://doi.org/10.1093/ccc/tcac039.

the precondition for a correct interpretation. To appropriate Nicholas Carr's famous question "Is Google making us stupid?," this argument would conclude that Netflix makes us stupid as well, as the smooth and flat leads to an ahistorical, and therefore lacking, cultural understanding. However, quoting Ani Maitra and Rey Chow's argument against the expression of "media in Asia," Thomas Lamarre sees the question of context as falling into the trap of a "methodological individualism," where platforms, cultures, creators, and audiences are all seen as separate individuals.[8] Instead, he argues for the concept of platformativity, which adopts an infra-individual, intra-acting view where context will be part of an ecology of human-technology-society. According to Lamarre, flattening is not always bad: "It may prove equalizing in the sense of challenging hierarchies and democratizing interactions or equalizing in the sense of rendering equivalent, transforming into exchange value."[9] In other words, the platform aesthetics of the smooth and flat might function as an equalizing force for East Asian drama series to gain international impact despite US cultural hegemony, even if their production and nonlocal reception may not provide the most sophisticated cultural understanding. The smooth and flat reducing cultural linguistic barriers might not be a bad thing as long as the smooth leaves space for contemplative moments of pausing or hesitation, whereas the flat functions as a surface with embedded hyperlinks that the audience, if intrigued while watching, can search and learn about the locally specific histories and additional context. Just like how the superflat movement can either celebrate or critically engage with the shallowness of consumer culture, Netflix shows like *Squid Game* can also function as a double-edged sword. Similar to *Parasite*, the show can be ironically interpreted as a celebration of capitalism by capitalizing on its anti-capitalist message. However, it also retains value in raising

---

8. Thomas Lamarre, "Platformativity: Media Studies, Area Studies," *Asiascape: Digital Asia* 4 (2017): 285–305.

9. Lamarre, "Platformativity," 302.

awareness and inspiring interests for a non-Euro-American-centric culture. In other words, although platform cosmopolitanism is ridden with the connotation of neoliberal capitalism, it could also be an opportunity for East Asian cultural products to go beyond national and regional borders to reach an impact that cinema and television could never give.

Yet, who is responsible for these attempts at countering the challenges of uneven globalization, the totalization of media's power and influence, the Orientalization of East Asian cultures, and the rather narcissistic and ahistorical nature of consumption from a foreign audience Inspired by Lamarre's idea of platformativity, platform cosmopolitanism might provide a fundamental basis for countering the totalizing and debilitating view of the inescapability of capitalism and new media technology. Platformativity addresses the entangled set of relations among the platform, culture, content, and human, comprising an encompassing media ecology. Building on this view, platform cosmopolitanism points to a hope for the human participants in this media ecology to counter the homogenizing effects of media technology and globalization in their ways of infra-individual intra-action.

While the global success of *Squid Game* benefits the Korean entertainment industry, it also provides an opportunity to expand their reach while integrating moments of reflection. The writers and directors can incorporate appropriate context without deviating too much from the intended story through smarter and subtler means, such as offhand conversations between minor characters, news broadcasts playing in the background, and indirect references from specific props carried by the characters (newspapers, books, posters, etc.). For the audience, although they are not required to learn more about the sociohistorical context while consuming foreign media, it is an ethical act to do so. Lastly, anglophone scholars of East Asian studies are well-positioned to share knowledge by completing more public writings. East Asian dramas are owned and produced by global streaming devices that are intended to be accessible without local or specific context. However, research-sharing can enable the viewer to have a multilayered experience, rendering the show (and any associated works) more impactful in the end. Such acts should not be viewed as the actions of excavating contextual value. Instead of economic terms, we can think of scholars as being a part of the media ecology, participating in infra-individual intra-actions through research-sharing.

It is possibly unrealistic to expect one show to change the world by overthrowing neoliberal capitalism and Euro-American-centrism. Still, content creators and viewers have some agency in the ecology of platform-content-human to maintain some profound aesthetic experience in the age of the smooth and flat. The delay, the pause, the aghast that give the audience a moment of reflection is usually associated with art—can it be provided by mass media too?

Review of *Light the Night*

SHUWEN YANG

Abstract

This review offers a brief analysis of *Light the Night* (2021), including its background, plot, and characters. It also discusses the history of Taiwanese drama and the collaboration between Netflix and Taiwanese talents in the TV industry.

Keywords: *Light the Night,* Taiwanese Drama, Netflix Platform, OTT Media Platform, Global and local

On October 6, 1988, on the east slope of Jinan Mountain in Zhongshan District, Taipei City, a group of young college students are out on a photography club activity. Suddenly, a female corpse appears in the shaky lens of their camera—it is in such shock and suspense that the story of *Light the Night* (2021) kicks off. As the plot unfolds, the audience is drawn into a web of mystery and intrigue as they discover that the murder is linked to a Japanese-style nightclub called Hikari, which means "light," where one of the hostesses is the likely victim.

Produced by Ruby Lin, directed by Lien Yi-chi, and written Ryan Tu, *Light the Night* is a Netflix original series comprised of three seasons, each with eight episodes. The first two seasons were released in November and December 2021, respectively, with the third season following in March 2022. The drama boasts an impressive cast, featuring stars such as Ruby Lin, Cheryl Yang, Tony Yang, Rhydian Vaughan, and Derek Chang. It delves into the complex relationships among the hostesses working at the Hikari nightclub, exploring themes of jealousy, heartbreak, friendship, love, and betrayal.

https://doi.org/10.3998/gs.4288

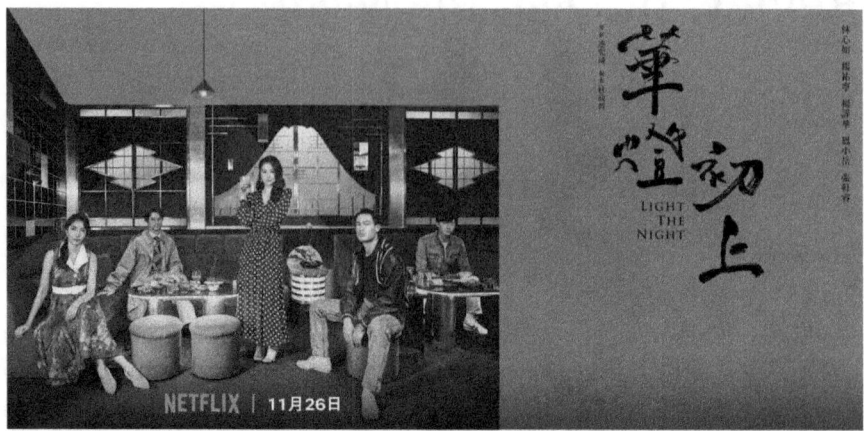

Figure 1: Netflix Poster of *Light the Night*
Source: Netflix Newsroom https://about.netflix.com/en/news/in-star-studded-light-the-night-a-unique-taipei-subculture-comes-alive

The central plot of the drama revolves around the murder-mystery surrounding the unidentified corpse. The first season depicts the story building up to the murder, with the victim's identity revealed in the concluding episode. The second season investigates the motive behind the crime, culminating in another murder and further heightening the suspense. The identity of the murderer remains a mystery until the end of the third season. The show's immense popularity can be attributed to its use of diverse genre elements such as romance, crime, and suspense, as well as its nonlinear narrational form, as seen in its numerous flashbacks and interludes. Based on Flixpatrol's data,[1] the drama secured the top spot in Netflix Taiwan's most-watched list on its second day of release (November 28, 2021). It maintained its position in the top ten for a total of 103 days. Additionally, the show also made it to the top ten in Netflix Hong Kong, Singapore, and Malaysia.

---

1. "Light the Night Top 10," Flixpatrol, accessed April 1, 2023, https://flixpatrol.com/title/light-the-night/top10/.

The story is set in the "Tiao Tong" (*jo dori*) area in Taiwan in the 1980s. "Tiao Tong" refers to "alley" in Taiwanese usage of Japanese, and the Tiao Tong area is comprised of nine parallel alleys near Taipei Main Station. Each alley is named from Tiao Tong One (*ichi jo dori*) to Tiao Tong Nine (*kyu jo dori*), and they are connected by the main road, Lin Sen North Road. In Taiwanese vernacular memory,[2] Lin Sen North Road was a synonym for red-light districts and prostitution in the 1980s. During the Japanese economic boom in 1980, many Japanese investors came to Taiwan, and their after-work drinking culture fuelled the growth of Japanese-style nightclubs. Additionally, on July 15, 1987, Taiwan lifted martial law, ushering in economic liberalization and consumerism, which led to an increase in locals' willingness to spend time in entertainment venues. Hikari nightclub was born out of this era. The hostesses at Hikari sell emotional companionship and romantic fantasies while striving to maintain their dignity. However, their relationships with the guests and with fellow hostesses inevitably evolve into complicated issues. Such a setting allows the drama to portray the stories of the marginalized group of hostesses, explore the existence and authenticity of "love," and engage a sense of nostalgic sentimentality.

Set against the backdrop of Tiao Tong, the drama's most prominent aspect lies in its portrayal of the female characters within the red-light district. The focus is on two female characters, Lo Yu-nung Rose (Ruby Lin) and Su Ching-yi Sue (Cheryl Yang), and their emotional entanglements over the years, presented through a nonlinear storytelling approach. As the story unfolds in reverse order, we witness the depth of their friendship as they navigate their past. Having been neighbours since high school, they have relied on each other through thick and thin. Su Ching-yi suffers from the trauma of being raped by her stepfather with her mother's acquiescence, followed by an unwanted pregnancy, which ultimately leads her to leave her family

---

2. Chen Yonghan 陳永翰, "Tiao Tong" "條通," Di Shiqi Jie Taibei Wenxuejiang Dejiang Zuopinji 第十七屆臺北文學獎得獎作品集 (The collection of winning works of the 17th Taibei Literature Award) (Taibei: Taibei Shizhengfu Wenhuaju, 2015), 110–15.

behind and turn to Lo Yu-nung for refuge. During the process, Lo comes to Su's aid with unwavering support. Meanwhile, Lo's life is almost equally tumultuous, as she abandons her college studies to marry her boyfriend Wu Shao-chiang, only to end up in prison for several years due to Wu's illegal activities in business. Ultimately, the two female characters both find solace and a new beginning at Hikari, where they become business partners under the pseudonyms of Rose and Sue. However, conflict takes place as they fall in love with the same man, screenplay writer Jiang Han. For all those years, Rose and Sue remain steadfast by each other's side, making their eventual separation all the more heartbreaking.

While the intricate entanglement between the two friends serves as the primary plotline of the series, the drama ultimately presents a collective portrayal of a multitude of unique female characters, including Hana, A Chi, Aiko, and Yuri serving as other hostesses at Hikari. Through the narration of their stories, the drama generates a sense of uncertainty regarding who the victim, perpetrator, and instigator truly are. Furthermore, this group portrayal also provides a vital opportunity to explore various dimensions of female trauma, allowing for a richer and more nuanced representation of women. As such, no single character can be deemed the protagonist, as they are all integral in crafting a complex and multilayered story.

Hana, one of the most tragic female characters in the series, is once coerced into prostitution by her ex-boyfriend and goes to jail for manslaughter. The depiction of Hana's story explores the use of film language to convey the brutality of sexual assault. Through the use of techniques such as montage, splicing the scene where Hana is raped in the rain with the scene when she takes a shower after returning home, as well as employing jump cuts between the scene of her manslaughter and the one where she repeats the mistake, the drama underscores the enduring impact of trauma by portraying the narrative itself as a scar of trauma. The character A Chi, who is initially considered the least popular figure at Hikari due to her gambling debts and frequent arguments with other hostesses, gradually evolves as we witness her dysfunctional family and her indifferent parents. By focusing on A Chi's story, the drama highlights the relationship between a woman and the space she occupies, as A Chi lacks "a room of her own" and is constantly confined to narrow spaces such as her bed or the backyard, signifying an oppressive confinement of women. Yuri, the cool and distant hostess, shields herself behind a defensive facade. The drama exposes her vulnerable side as she seeks solace in the company of a host named Henry but unfortunately becomes involved with a drug-dealing group because of him. The drama aims to create multifaceted and diverse characters with the women of Hikari. Loneliness, the fragility of human relationships, and the deep yearning for love are recurrent motifs throughout their lives.

Mutual salvation is undoubtedly one of the profound meanings of the word Hikari. The story, overall, tells how female characters bond with, support, and help one another. Since the moment Rose and Hana become friends in jail, Rose has tried her best to protect Hana and help her rebuild herself after a traumatic experience. As the story progresses, a shift also happens in A Chi, who gradually starts to take responsibility for other hostesses as well as the club. Near the end of the third season, A Chi finds out she is pregnant and decides to leave. Rose goes to her place, and in their conversation, Rose points out, "The place where you do whatever you want, where you are truly yourself, is where you call home." When A Chi, the most "annoying" hostess, announces her pregnancy, the characters each uniquely show that they care. Such female bonding is perhaps the central meaning of "light."

Despite the deep exploration of female bonding, the audience may question how such profound friendships, especially between Rose and Sue, could crumble due to a man's involvement. The drama attempts to address this by offering an alternative explanation for female jealousy, which is also the reason behind the transition in Sue's character. Though she initially appears to be the perfect hostess, Sue harbors a desire for revenge against Rose before her departure from Taiwan for Japan, a revelation that shocks and confounds many viewers. In her diary, she describes the pain she has been feeling due to Rose's patronizing behavior, a sentiment that other female characters like A Chi and Hana share. Their repeated question "How could you patronize me?" highlights the underlying resentment in their relationships.

Jealousy, a trait often conflated with womenhood, has been traditionally depicted in Chinese culture as a rivalry between wives and concubines. In the first season of *Light the Night*, Sue's jealousy toward Rose initially seems to stem from a competition over a man. However, the series delves deeper into this theme, exploring women's pursuit of self-establishment and their quest for agency. Sue's suffering arises from being constantly "saved" but never taking control of her own destiny, taking responsibility for her child, or having a say in her love for Jiang Han. It is also noteworthy that the female characters at Hikari are identified mainly by their Japanese pseudonyms, which obscures their true identities and indicates a lack of control of their lives. Trapped by their names and by their profession, where emotions are feigned, their jealousy and resentment toward Rose reflect a broader sense of helplessness and lack of agency that extends beyond the traditional definition of "female jealousy." It is their way of striving for a voice and asserting their individuality.

In addition to the complex relationships among the female characters, *Light the Night* incorporates various genre elements such as drug dealing, violence, corruption, crime, and gangsters, which add to its appeal and commercial success. However, the inclusion of these elements also presents challenges, such as the occasional repetition of information from the first season, which can slow the pace of the later episodes. Nonetheless, the drama's release over a span of three months compensates for this flaw. Furthermore, the genre mixture allows for the inclusion of diverse and dynamic characters, such as the male characters Pan Wen-cheng (Yo Yang), a policeman, and He Yu-en (Derek Chang), a college student who later becomes an intern journalist. These characters complicate the plot and draw attention to issues like corruption and journalistic ethics. Another example of the show's diverse characters is the drag queen played by Wu Kang-ren, who challenges societal norms with seemingly hilarious questions like, "Can't a man be a hostess?"

Although only comical at first sight, the later development of this character indeed brings a bold and refreshing perspective to the drama.

The richness of the drama is also evident in the details. The drama incorporates allusions to popular songs that prompt us to consider the political climate of the story. The 1980s were a decade marked by a series of significant events in Taiwan. Martial law was imposed from May 20, 1949, to July 14, 1987, creating an atmosphere of tension and terror. Popular songs were banned due to their potential political implications and news was strictly censored. The drama seeks to incorporate cultural elements by featuring the recurring Taiwanese song "May You Have Happiness," initially released in 1972 and later celebrated for its empathy toward ordinary people like Hana. The drama also challenges cultural hierarchies. Hana's lack of Japanese language skills and her rural background initially prevent her from joining Hikari. However, everyone eventually appreciates her singing, reflecting an attempt to value Taiwanese culture. Also, He Yu-en's character, examined against the backdrop of the lifting of martial law, should not be seen as a naive college boy falling in love with a hostess but rather as a young man who later enters the news industry and seeks the truth. The fact that Rose has the opportunity to address the media and explain the circumstances surrounding the murder further illustrates the drama's attention on the news industry.

The endeavour of *Light the Night* to tackle comprehensive and sophisticated themes should not be viewed as isolated. Instead, it warrants examination within a broader context, offering insights into the historical development of Taiwanese dramas. At the beginning of the twenty-first century, Taiwanese dramas were characterized by the "idol soap opera" style, which featured simplified and unrealistic romantic love stories between young men and women, following the tradition of Qiong Yao dramas from the 1950s and 1960s. Famous dramas include *Meteor Garden* (2001), *It Started with a Kiss* (2005), *The Prince Who Turns into a Frog* (2005), and *Fated to Love You* (2008), which were popular in Taiwan and have been imported into mainland China. However, with *In Time with You* (2011) as the division point, Taiwanese dramas experienced a decline around the year 2011.

Facing such a decline, there have been many efforts to explore a way out. A significant event was the emergence of a new form of TV series called Qseries, 植剧场, in 2016. The character "植" literally means "plant," which, according to chief inspector Wang Shaudi, signifies the effort to root TV drama in Taiwan and cultivate new talents for the TV industry.[3] The series comprises eight parts, with a total of fifty-two episodes, covering four major genres: love and growth, suspense and deductive, supernatural and horror, and adaptation from original works. In the 2017 Taiwan Golden Bell Awards, the Best Drama Award was given to two dramas coming from Qseries: *Close Your Eyes Before Its Dark* (2016) and *A Boy Named Flora A* (2017). Such exploration of different genres is paving the way for the future production of genre-mixing commercial dramas.

The transformation of the Taiwanese TV industry was widespread, with numerous individuals switching roles. Ruby Lin, once the leading actress in Qiong Yao-style dramas, has become a producer. Similarly, Alyssa Chia, also a well-known actress in idol soap dramas, produced the critically acclaimed drama series *The World Between Us* (2019), a Netflix original series that achieved phenomenal success in 2019. In an interview, Lin talks about how, as an actress, there are certain characters such as hostesses that she wants to explore, and emphasizes the necessity for workers in the Taiwanese drama industry to evolve and adjust to new roles and performances in response to the genre's diverse demands.[4]

---

3. Haofengguang Chuangyi Zhixing好風光創意執行 and Yuandongli wenhua原動力文化, Yichang wenrou Geming: Zhijuchang quanjilu一場溫柔革命：植劇場全記錄. (A gentle revolution: documentation of Qseries) (Taibei: Yuandongli Wenhu Shiye Youxiangongsi, 2018).

4. Fei Yu 飞鱼, "Huadeng Chushang: Sheiren Buzai Huanle Chang, zhuanfang lin xinru 《华灯初上》：谁人不在欢乐场 | 专访林心如" (Light the night: Who is not in the entertaining circle, an exclusive interview with Ruby Lin), Toutiao.com, accessed February 10, 2023, https://www.toutiao.com/article/7083134824801960484/?source=seo_tt_juhe&wid=1678405646200.

Furthermore, such development not only highlights the increased participation of female figures in the drama-production industry but also demonstrates how overseas over-the-top (OTT) media-service platforms like HBO, FOX, and Netflix have begun to produce Asian TV dramas, sparking a revival of Taiwanese dramas. In a recent interview with "Yu Li,"[5] Chih-Han Chen, the producer of *Someday or One Day* (2019), discussed the three primary approaches to producing Chinese dramas for overseas OTT platforms. The first method involves full investment and high involvement from the OTT platform, with local production companies only responsible for production. The second approach is a joint investment between the OTT platform and the film and television company, with the former buying out the distribution rights of the project. In this scenario, the script has already been completed when communication and cooperation begin. The third approach is a collaboration between TV stations and film and television companies, with the platform deciding whether to cooperate after seeing the finished drama. According to Lin,[6] this is the case for *Light the Night*, so the production team gets to keep all the content from the beginning. The extent to which OTT platforms should be involved in content development and how to achieve a balance between local and global/universal elements remain unresolved issues, as we see that adjustments are constantly made in different methods and cases.

Nevertheless, *Light the Night* is certainly a successful production of such cooperation. With its unique setting in 1980s Taiwan, the remarkable performances by Taiwanese actors and actresses, and Netflix's professional marketing strategies, the series has successfully found a balance

---

5. "Taiju Ruhe zai Liangniannei Wancheng Nixi: Chen Zhihan zhuanfang 台剧如何在两年内完成逆袭？陈芷涵专访" (How did Taiwanese drama revive in two years? An exclusive interview with Chen Zhihan), k.sina.com, accessed February 10, 2023, https://k.sina.com.cn/article_5737990122_15602c7ea01900lkys.html?cre=tianyi&mod=pcpager_focus&loc=18&r=9&rfunc=100&tj=none&tr=9.
6. Fei Yu, "Interview with Ruby Lin."

between entertainment and popularity and seriousness and social responsibility. Netflix commented on Taiwanese drama that "Southeast Asia is a large region with diverse and complex cultures, and Taiwanese dramas provide unique stories."[7] How will OTT platforms depict stories within the Sinitic-language world in the future? How can Taiwan further contribute to the global stage? *Light the Night* is just the beginning, shedding light on these questions that remain to be explored.

---

7. "Sandairen de qingchun: liushinian taiju xingshuaishi 三代人的青春：六十年台剧兴衰史" (The youth of three generations: The rise and fall of Taiwanese Drama in sixty years), Everyone Focus, accessed February 10, 2023, https://ppfocus.com/0/he631bcbd.html.

Short Essay

Three *Bad Kids*, One Loving Killer

Red China Noir in Blakean Symmetry

SHENG-MEI MA

Abstract

The twelve-episode Chinese TV series *Yinmi de Jiaoluo* (隱秘的角落*Hidden Corner*, translated as *The Bad Kids* [2020]) dances on a Blakean "The Tyger and the Little Lamb" tightrope between childlike innocence and homicidal nihilism, between an art house sensibility and a pop culture chained to party propaganda. Amidst the flood of ethnocentric and jingoistic police procedurals "with Chinese characteristics" on TV, director Xin Shuang (辛爽) energizes his tour de force with a sensibility ranging far beyond Chinese shores, flirting with Western artists and metaphysical self-reflexiveness torn between good and evil, innocence and meaninglessness. Xin Shuang adapts Zijin Chen's (紫金陈) eponymous web novel while imbuing the series with an off-kilter, haunting Yeatsian "terrible beauty" of violence and attraction.[1] *The Bad Kids* made a killing not so much in profits as in the true art of Sino Noir, or Red China Noir.

The eponymous "bad kids" blackmail a murderer to obtain funds for a life-saving surgery. Courting his own death, this "loving" killer saves one of the three kids from an asthma attack and spares the other two out of a fatherly compulsion to sire his own offspring, to pass on the legacy of revenge and guilt, to prolong his life—his afterlife, rather—as he confides: "I want you all to live—to live like me." *The Bad Kids*' Red China Noir teeters on a Blakean symmetry of love and hate, East and West.

---

1. Zijin Chen (紫金陈). *Huai Xiaohai* (壞小孩; Bad kids). Hunan: Hunan Wenyi, 2014. https://www.99csw.com/book/5956/index.htm.

Keywords: *Yinmi de Jiaoluo,* The Bad Kids, Xin Shuang, Qin Hao, Red China Noir

The twelve-episode Chinese TV series *Yinmi de Jiaoluo* (*Hidden Corner*, translated as *The Bad Kids* [2020]) dances on a Blakean "The Tyger and the Little Lamb" tightrope between childlike innocence and homicidal nihilism,[2] between an art house sensibility and a pop culture chained to party propaganda.[3] The suffix of "the" in *The Bad Kids* is grammatically redundant in the Chinese language, which has no definite articles. Conceivably, *The Bad Kids* can be shortened into *Bad Kids*, a perfect rendition of the web novel title *Huai Xiaohai*, from which the TV series is adapted. This bilingual backstory leads to the titular liberty of "Three *Bad Kids*," alliterating with the nursery rhyme "Three Blind Mice," dogging and dogged by one killer cat, so to speak.

As the "Chinese Century" uncoils amidst global unrest and the pandemic of COVID-19, as China enjoys a comparatively low crime rate vis-a-vis industrialized and industrializing nations, *The Bad Kids* is but one show out of what amounts to a "crime wave" in TV series. This growing collective (collectivist?) fascination with crime suggests paranoia over the West's capitalist sins of social inequality and alienation; over China's own vices censored, repressed, and even institutionalized, normalized; or over both. Apropos the virus infecting the "good" China, does it hail from the "bad" West, or is it endemic to China itself? Per W. B. Yeats's "The Second Coming" (1920), is the beast slouching from Bethlehem toward Beijing, or is Beijing the beast? On the one hand, crime shows intimate these unsettling doubts on viral evil, a theoretical, theological premise out of keeping with secular, pragmatic Confucian-cum-communist-materialist ideology. On the

---

2. William Blake, *Songs of Innocence and of Experience*, 1789–94, Project Gutenberg, accessed May 16, 2023, https://www.gutenberg.org/files/1934/1934-h/1934-h.htm.

3. Xin Shuang, dir., *The Bad Kids* (隱秘的角落), performances by Qin Hao and Wang Shengdi, iQIYI, 2020.

other, crime shows expound law and order exemplified by the public security known as *gongan*, equivalent to the police in the West, not only to toe the party line but also to alleviate, subconsciously, existential angst. Amidst such ethnocentric and jingoistic police procedurals "with Chinese characteristics," director Xin Shuang (辛爽) energizes his tour de force with a sensibility ranging far beyond Chinese shores, flirting with Western artists and metaphysical self-reflexiveness torn between good and evil, innocence and meaninglessness. *The Bad Kids* made a killing not so much in profits as in the true art of Sino Noir, or Red China Noir.[4]

Xin Shuang adapts Zijin Chen's (紫金陈) eponymous web novel while imbuing the series with an off-kilter, haunting Yeatsian "terrible beauty" of violence and attraction.[5] From the outset of the opening credits, Xin Shuang couples an Escheresque Möbius strip of stairwells with an eerie, disorienting, and off-key soundtrack accompanying three white will-o'-the-wisps chased by a looming blob of darkness in a nightmarish game of peekaboo or cat-and-mouse (Figure 1). The three flitting phantoms represent the titular "bad kids" while translating, literally, the common expression *xiaogui* for "little ghosts" or "little kids." The soundtrack music punctuates a male or female vocal gasping, perhaps expiring. This disjointedness intensifies in the crime accidentally witnessed by the three adolescent protagonists, all from broken homes, one excellent student Zhu Chaoyang and two orphans Yan Liang (Ding Hao in Zijin Chen's web novel) and Pupu. The first name Liang means "good." Pupu, notwithstanding the English soundalikes of "poo(h)," means ordinary, common, drawing viewers close to the lovely, petite "everygirl" à la the medieval morality play *Everyman*. Despite her young age, Pupu's elocution is so affective that she delivers some of the most philosophical reflections: "Have you studied hard so that you can kill?" she queries

---

4. A portmanteau of the communist "Red China" and the genre of "film noir," Red China Noir points to representations of crime and detection "with Chinese characteristics," or within the context of the communist political, social, and cultural apparatuses.
5. W. B. Yeats, "The Second Coming," *The Collected Poems of W. B. Yeats* (New York: Macmillan, 1959), 184–85.

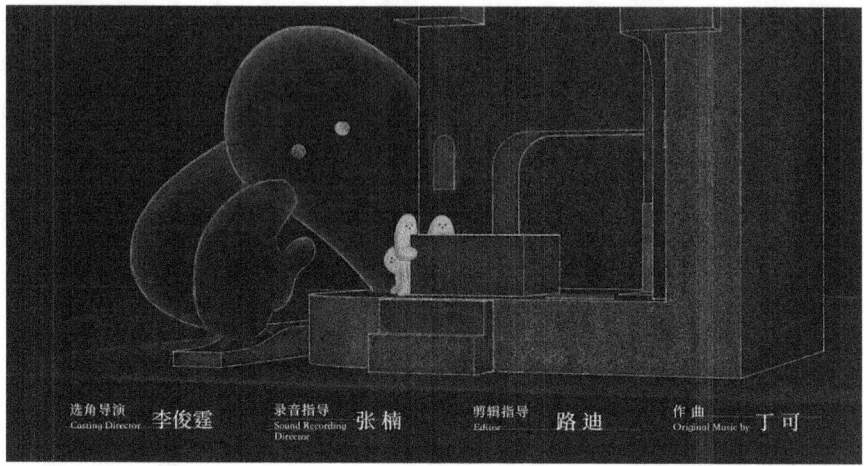

Figure 1: Three little ghosts chased by a big one in the opening credits to
The Bad Kids.
Source: A screenshot from *The Bad Kids*, subject to the fair use rule in
copyright laws

the teacher-murderer; "Are criminals always criminals?" She queries her two
"partners in crime." Both are rhetorical questions to a certain extent, yet it
remains an open question whether her elders would answer in the affirmative or in the negative.

The orphans run away from the orphanage to try to scrounge up money
to pay for a life-saving surgery for Pupu's younger brother. As they video
themselves on their rare outing, singing the children's song *Xiao Baichuan*
(literally "little white boat") together at a tourist site, they capture unwit-
tingly the murderer Zhang Dongsheng shoving his parents-in-law over an
adjacent peak. The middle-schooler's first name Chaoyang (morning/rising
sun) eerily repeats the murderer's name Dongsheng (east rise). The symmetry of innocence and evil hides in plain sight of the characters' names so
common as to go unnoticed by censors until one realizes that Chaoyang
is a key government and financial district in the capital Beijing. "Dongsheng," more problematically, evokes not only Chairman Mao Zedong, the

proverbial Red Sun, but also the revolutionary song forever associated with Mao. "The east is red. The sun rises," the tune goes: "China has sired a Mao Zedong." To name a fictional murderer after a historical mastermind of famines and "reeducation" pogroms seems apt, but it is tantamount to suicide under Xi Jinping modeling himself after the father of Chinese Communism.

Zhang commits the double murders to win back his wife who is seeking a divorce under her parents' "bad influence." Instead of turning over the evidence to the police, the kids fancy extorting money from Zhang in hope of financing the surgery. An echo of the opening credits' stairwells that drop off and the soundtrack that breaks off, the bad blackmailers' good intentions set in motion a chain reaction of deceptions, police fumbles, involuntary violence, and four more deaths: Zhu's spoiled, willful half-sister; his hysterical stepmother; his grieving father; and the half-sister's doting gangster uncle. Courting his own death, the "loving" killer Zhang is shot and killed, after having saved Pupu from an asthma attack and spared Yan Liang's and Zhu's lives out of a fatherly compulsion to sire his own offspring, to pass on the legacy of revenge and guilt. Pupu's bout of asthma was so severe that she had lost consciousness. When Zhang called the ambulance, he did so at great risk to himself as a suspect and fugitive.

The Bad Kids' Red China Noir teeters on a Blakean symmetry of love and hate, East and West. Such Blakean symmetry is as much complementary as it is contradictory. Out of an obsessively pathological love, Zhang disposes of his parents-in-law to cling onto his wife. Initially, Zhang takes two lives to continue his abject, masochistic enslavement to a wife who despises him. Subsequently, Zhang is compelled to kill again and again to conceal his crime. Out of love for Zhu, Pupu corners Zhu's bully of a half-sister, indirectly causing the frantic girl's fall from the school building. Although Pupu harbors no evil intent like Zhang's, the result of a smashed body looks similar to Zhang's premedicated murders. Out of the urge to prolong his life—his afterlife, rather—Zhang refrains from knifing Yan Liang, confiding: "I want you all to live—to live like me."

Zhang's exhortation resonates with the serpent's seduction of Eve in the Book of Genesis and Mephistopheles's bargain with Faust in Goethe.[6] The drive for self-aggrandizement is shared by God, who shows off Job—"Hast Thou [Satan] considered my servant Job . . . a perfect and upright man?"[7] (*The Book of Job* 1:8)—and by Satan, who deceives, tactically—"If what is evil/Be real, why not known, since easier shunned?"[8] To mimic God's omnipotence, Satan plots to have evil, once known, be shared and not shunned, hence enlarging the Satanic Kingdom on Earth. Nietzsche's Zarathustra, of course, dismisses both the biblical and Miltonic fallacy, centering good and evil squarely in the human heart that wields the knife and hatches the serpent: "His soul wanted blood, not booty: he thirsted for the happiness of the knife!," and: "What is this man? A coil of wild serpents that are seldom at peace among themselves—so they go forth apart and seek prey in the world."[9] Akin to Eve, Faust, and the "coil of wild serpents," all three kids are to live on, allegorically, as Zhang's "bad" children, to bear Cain's mark of sin—not the Semitic fratricide against Abel but the Sinitic pseudopatricide against Zhang. The childless Zhang also feigns stabbing Zhu to force the police to shoot him. In the last episode, his last words to Zhu come as a parting shot at the kids and the world. Witnessing Zhang's knife raised high in the air, the police resort to deadly force to save Zhu from what they believe to be mortal danger. Zhang is slain midsentence: "You can believe in fairy tales . . . " (Figure 2).

The blood on Zhang's left shoulder in Figure 2 comes from Zhu's stab wound earlier, not the gunshot about to bring him down. Goading Zhu to avenge three lives—his father, whom Zhang killed with a screwdriver, and his friends Yan Liang and Pupu, whom he let go or saved without Zhu's

---

6. Johann Wolfgang von Goethe, *Faust*, translated by Walter Arndt (New York: W. W. Norton, 1976).

7. Book of Job 1:8.

8. John Milton, *Paradise Lost*, in *The Complete Poetry of John Milton* [[editor?]] (New York: Anchor, [1667] 1971), 698–99

9. Friedrich Nietzsche, *Thus Spoke Zarathustra: A Book for All and None*, 1883 and 1885, Project Gutenberg, accessed May 16, 2023, https://www.gutenberg.org/files/1998/1998-h/1998-h.htm#link2H_4_0011.

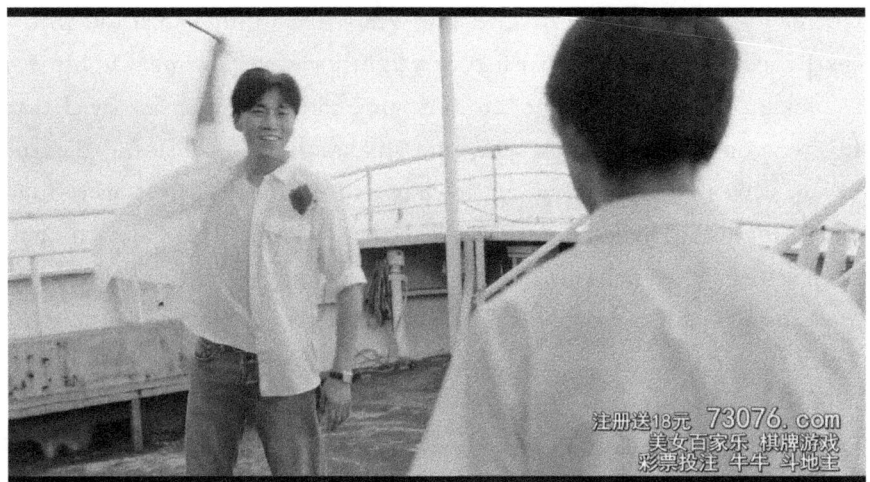

Figure 2: The murderer Zhang Dongsheng about to be slain in episode 12 of *The Bad Kids*.
Source: A screenshot from *The Bad Kids*, subject to the fair use rule in copyright laws

knowledge—Zhang tempts Zhu to follow his own path of violence, to become, as Yan Liang warned, "the second Zhang Dongsheng." If love and its fruits of children are out of reach, hate satiates the human need just as well for physical contacts through clashes of bodies. The fruits of hate hemorrhage into rotted body parts and tormented souls, both in denial of the finality of death. Maggots emerge from wounds, murders from a diseased heart—both filling the void of mortality.

With a pleasant smile, Zhang dangles midair his weapon of choice to invite the bullet. Zhang's open, welcoming body language reminds one of Sigmund Freud's "bad" children misbehaving, "to provoke a punishment of some kind, and that after they have been punished they calm down and are quite happy."[10] Failing to clone a second Zhang Dongsheng, he

10. Sigmund Freud, "Criminals who Act Out of a Consciousness of Guilt," in *The "Wolfman" and Other Cases*, translated by Louise Adey Huish (New York: Penguin, [1916] 2003), 346–48.

clowns to hasten his own demise. His weapon is a round steel rod with a sharpened end; its handle consists of a tightly wrapped twine. On his way to pick up the extorted money from Zhang, Yan Liang picks up by chance this weapon in his and Pupu's hideout, an abandoned boat stranded at the beach. This strange-looking "knife" has a fortuitous feel to it, as though Yan Liang improvises rather than proceeds by design. This slapdash quality permeates the heinous killings as well. Zhang's executions of crime look almost haphazard, knifing Zhu's father with a screwdriver, strangling Zhu's stepmother with a crowbar—both objects happen to be at hand. Even the final push over the cliff is done with his bare hands, after one final plea to the in-laws falls on deaf ears. "Impulsive crimes of passion," a seasoned defense attorney would no doubt argue in court, "not premedicated homicides." Akin to his in-laws' fall, Zhang's own fall into evil appears more accidental than intentional, making him somewhat sympathetic. Viewers may even share Zhang's fear as the kids and the police tighten the noose. The identification with the murderer, an empathy somewhat misplaced, stems from human nature, mirroring theatergoers' fright, alongside Macbeth's, over the knocking in the wake of Duncan's murder, or moviegoers', alongside Peter Lorre's in *M* (1931).[11] Thomas De Quincey's 1823 essay "On the Knocking at the Gate in *Macbeth*"[12] analyzes readers' projection onto the perpetrator of a regicide, the tenor of romantic self-inflation and overreach quite applicable to murders in Red China Noir two centuries later.

On the part of the "bad kids," the entire blackmail proceeds less according to plan than dictated by circumstances, once the three kids are

---

11. William Shakespeare, *The Tragedy of Macbeth*, in *Shakespeare: The Complete Works*, ed. G. B. Harrison (New York: Harcourt, Brace and World, 1971), 1184–1219; Fritz Lang, dir., *M.*, performances by Peter Lorre, Ellen Widmann, and Inge Landgut (Berlin: Nero-Film AG, 1931)

12. Thomas De Quincey, "On the Knocking at the Gate in *Macbeth*," in *English Literature Anthology for Chinese Students*, ed. John J. Deeney, Yen Yuan-shu, and Chi Ch'iu-lang (Taipei: Hongdao Publisher, [1823] 1978), 403–5.

taken hold by the urge to help Pupu's brother. Saving one child entails ending the lives of three adults and one child. Making it up as they go, Yan Liang happens upon the steel rod in a boat, secreting it in his school bag. In a series of botched confrontations, the rod passes from Yan Liang's school bag to a ploy in Zhang's hand to induce triple births of his very own "bad kids." Baited by Zhang "to avenge your father," Zhu thrusts the rod into Zhang inches above the heart. A lucky miss or a bad kid's inner goodness averting the point of the blade? Either scenario lands the rod right back in Zhang's hand, symbolizing the will to power over lives, even Zhang's own if not Zhu's. The rod embodies as much Yan Liang's instinct for self-preservation against Zhang as Zhang's instinct of self-regeneration out of Zhu, the heir apparent. Absent the clone, Zhang wields the rod in gleeful anticipation of self-annihilation. The phallic symbol of the rod delivers death—both homicide and suicide. Inadvertently, however, it also underwrites life with Zhang's blood, ensuring the bad kids' new lease on life with Zhu's single mother, Yan Liang's adoptive parents, and Pupu's and her brother's recoveries.

Given the Blakean symmetry of self and other, of human-ness and thingy-ness, the weapon of choice may have chosen its handlers in a post-Anthropocene, posthuman era. Yan Liang chances upon the rod in the same manner as he chances upon the hideout of a deserted boat, which is predetermined by the opening credits' white boat imagery (Figure 3). Although such metaphors spring from Xin Shuang's head, that head is created by a lifetime of experiences on Earth. Chicken or egg: which—the man or the world—is the creator, which the creature? In its making, the steel rod resembles an acupuncture needle, which is but steel wire sharpened at one end and wound tightly with more wire at the other end to form the handle. Theoretically, such a needle punctures skin to reach an acupoint, a node of qi or energy, to heal. It inflicts local pain to ease holistic pain. The steel rod kills and heals, just as Red China Noir elicits and exorcizes fears. The rod also conjures up the Buddhist vajra of a thunderbolt or a diamond sword of wisdom, shearing through illusions, laying waste to

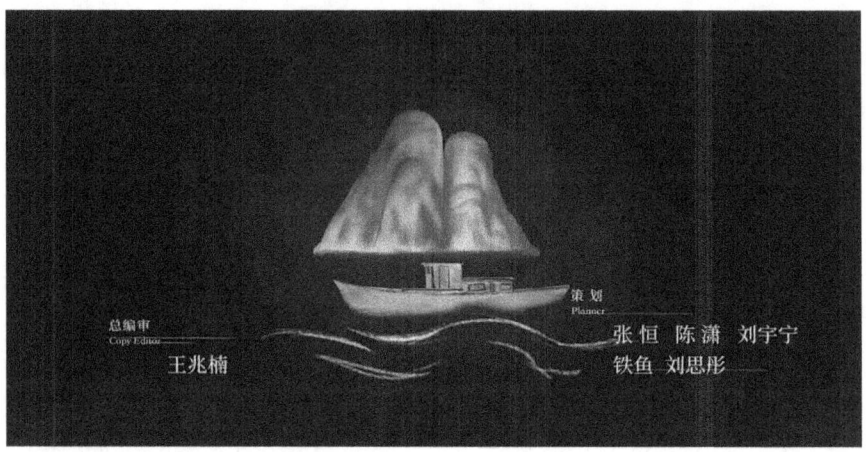

Figure 3: The white boat in the opening credits.
Source: A screenshot from *The Bad Kids*, subject to the fair use rule in copyright laws

dreams, for awakening to nirvana. Despite symbolism of traditional medicine and Buddhism, Zhang remains a "butcher" against others and the self, arrested in the first half of the popular Buddhist maxim: "Lay down the butcher's knife, turn into a Buddha anon." Zhang's stunted growth, refusing to be "weaned" from the imago of a spouse, provides a wry contrast to childhood innocence.

Zhang's last words on fairy tales loop back to, ironically, the recurring motif of the Cartesian coordinate system mused, respectively, by Zhu the math wizard and by Zhang the substitute math teacher at a cram school. In their shared rumination over Descartes's creation and love life, they actually think alike, almost a father-son, master-disciple pair. A coordinate system comprises two intersecting perpendicular and horizontal lines, each marked by positive and negative coordinates from the origin or the intersecting point. To situate *The Bad Kids*, intersecting coordinates are essential, juxtaposing East versus West, art versus politics, noir versus red, life versus death, and romantic make-believe versus reality. Legend has it, and a twice-told tale from Zhu and Zhang at that, that Descartes served as the Swedish queen Christina's philosophy tutor and fell in love with her thirty years his junior. Barred from marriage by the court, Descartes died of pneumonia and a broken heart. A fairy tale with a tragic ending, the reality, as Zhang relates it, is that the queen never cared for the lovelorn Descartes, who rose early in freezing Scandinavia to instruct the queen on the philosophy of love, contracting pneumonia as a result. Zhang's last words begin by confirming fairy tales of unrequited love, yet this collective wish fulfillment is cut short twice: by a bullet on the abandoned boat and by a bug by the Nordic Sea, both seized midway through the father-lovers' confessions.

The ruin of a boat by the sea returns to the "first impression" of *The Bad Kids* or Figure 3: the opening credits' ink-brushed or computer-generated boat, cruising and morphing into three white will-o'-the-wisps haunted by a (their own?) long black shadow. Shaped like an uroboros biting its own tail, the series ends up circling back to where it started. The closing credits to the finale, episode 12, unfurl under the intertitle of "Dedicated to Childhood," a serial collage of numerous actors' and crew members' childhood photographs, some pictures going as far back as the Cultural Revolution. This tribute favors, apparently, more senior team members, thus containing no baby picture for the main cast's youngest, the barely ten-year-old Wang Shengdi, who plays Pupu. This moving, bittersweet closure may well be an epitaph to the futility of existence, since the visual pathos accrued from innocent children and infants is montaged, auditorily, by the two-part theme song of "White Boat" and "Little White Boat," delivered in tandem by Qin Hao, who plays Zhang Dongsheng, and by Wang Shengdi, as much a "tyger" and "little lamb" pair as they come. While the viewer's eyes mist over, stirred by the melancholia of time past and paradise lost, their ears are shocked by Qin Hao's riff of "White Boat" sans "Little" or innocence, de facto Xin Shuang's final words, who revises the 1924 Korean song "Little White Boat" by Yin Kerong. Xin Shuang's "White Boat" runs as follows (Translation mine, no punctuation added for there is none in the original. A grammarian would feel compelled to "correct" the song title as "*The* White Boat," which defeats the whole purpose of dropping "little," divesting naïveté. Ditto the grammarian's standardizing of "Little White Boat" as "*The* Little White Boat"):

> When you fly
> The silvery white sail
> Sink into the bottom of the sea with the setting sun
> What you once owned
> Youth and restlessness
> No one would ever bother to bring up
> The naïve folk song
> Sung for whom
> You ingénues
> Where have you gone
> Just follow the waves
> Follow them
> There is no "the Other Shore" anyway
> For anyone to arrive[13]

Notwithstanding political correctness of an intertitular intrusion, mid-song, on the Communist Party's ongoing "Laws for the Protection of Youth" (Figure 4), "White Boat" is heart-wrenchingly bleak. "White Boat" in effect obliterates the Buddhist utopia beyond life's "bitter sea," beckoning from the other shore of blessedness. Qin's song of experience is then reprised in Wang Shengdi's angelic voice of the uplifting "Little White Boat," which, twelve episodes ago, has already accompanied two bodies

---

13. "Hidden Corner Episode 1," dramasq.com, accessed June 3, 2023, https://dramasq. com/cn200616/1.html#5; or IQIYI, "The Bad Kids EP01 | 隐秘的角落 | Qin Hao 秦昊，Wang Jingchun 王景春，Rong Zishan 荣梓杉," YouTube, accessed June 3, 2023, https://www.youtube.com/watch?v=XkQg0r93-eU (YouTube with English subtitles).

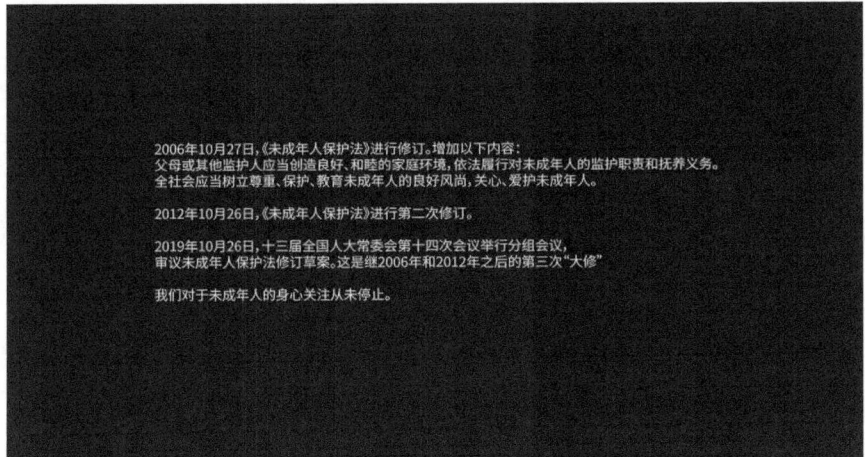

Figure 4: The intertitle on "Laws for the Protection of Youth" halfway through the song of "White Boat" in the closing credits of episode 12. *Source*: A screenshot from *The Bad Kids*, subject to the fair use rule in copyright laws

falling off the cliff. The children's earlier chorus of a flight of fancy punctuates the nosedive of two victims and the singers themselves. A Second Coming after Qin within the "family album" of "Dedicated to Childhood," Wang comes across as a spinoff of Qin's "parent" lyrics, although her words, as composed by Yin Kerong in Korean and translated into Chinese in the 1950s, predate Qin's. Whereas any song of innocence precedes, theoretically, the song of experience, the reversed order of lyrics puts adulthood ahead of childhood, echoing the sequence of the nightmarish series on Red China Noir before the nostalgic closer via a sepia-toned album of the Gilded Age long faded.

Wang delivers the Korean songwriter's lyrics, which mashes three Chinese legends of the moon, interlacing immortality with Tantalus's exhaustion and thirst in Greek mythology. The trio comprise of Chang'e, Wu Gang, and the Jade Rabbit, all tasked with perpetual repetitions that bespeak longings never fulfilled. Chang'e flies to the moon after swallowing the elixir of longevity, forever regretting the betrayal of her husband.[14] Wu Gang fells the *guihua* or osmanthus tree on the moon, only to have it grow back time and time again. The Jade Rabbit pestles immortality pills in the mortar by the tree, but the fact that she does so night after night begs the question of whether we on Earth look up at a lunar shadow play of transcendence or of punishment for millennia. The piling on of "bedtime stories" comes as though the Wordsworthian or Xin Shuangian Child—"Father of the Man" in "My Heart Leaps Up"—is trying to convince herself, to lull herself to sleep, a suspicion crystalized in the glitch of a line: "Oars, oars, seeing none/ Nor does the boat have any sail." Per Zhang's counsel, one is free to "believe in fairy tales."[15] The moon-bound Flying Dutchman, a ghost ship, requires no such earthly tools for flight, while the counselor's tool, his half-sword of half-wisdom, clink-clanks onto the deck as he drops dead.

Fairy tales are what the Sand-Man Xi Jinping's "China Dream" is made of.[16] Holding melodramatic soaps in one hand and triumphal crime-fighting police shows in the other, the millennial China Dream in popular TV series sprinkles sand in viewers' eyes, lulling them to a consensual, collective sleep, albeit too shallow and disturbed for some. Within Lu Xun's iron house of sleepers,[17] Xin Shuang plays on the porousness of reveries and wokeness,

---

14. Hou Yi is the husband of Chang'e, a great archer who shot down nine of the ten suns in the sky to save humanity from drought and desiccation. In return, Hou Yi is rewarded with the elixir of immortality, which Chang'e swallows on the sly and flies to the moon to opine her betrayal of Hou Yi.

15. William Wordsworth, "My Heart Leaps Up," 1802, Poetry, accessed May 16, 2023, https://www.poetry.com/poem/42279/my-heart-leaps-up.

16. E. T. A. Hoffmann, "The Sand-Man," in *The Best Tales of Hoffmann*, ed. with an intro. E. F. Bleiler (Mineola, NY: Dover Publications, [1817] 1967), 232–66.

17. The Father of modern Chinese literature Lu Xun deploys a pivotal metaphor in "Preface to *A Call to Arms*" (1922). Lu Xun imagines the turn-of-the-last-century China as "an iron house without windows" about to suffocate sleepers within. Crying aloud would awaken "a few of the light sleepers" only to intensify their "agony of irrevocable death" (5). Evidently, Lu Xun chooses to cry out nonetheless, inspiring Leo Ou-fan Lee's *Voices from the Iron House: A Study of Lu Xun* (Bloomington, Indiana: Indiana University Press, 1987). See Lu Xun. "Preface to *A Call to Arms*," in *Selected Stories of Lu Hsun*, trans. Yang Hsien-yi and Gladys Yang (Peking: Foreign Languages Press, [1922] 1972), 4–7.

of China and the West, of wakened and slumbering Chinese, just as the revisionist lyrics of "White Boat" are truncated and made right (left?) by a Model Peking Opera-style banner of an intertitle on youth-protecting laws. Amidst digitally streaming dreams to zombify a citizenry, *The Bad Kids* stands alone as a crypto-noir that secretes Red China's wake, a well-wrought funeral wake to sleep on, to mull over the deadening of viewing subjects and their immanent quickening.

CONTRIBUTORS

Yucong Hao is a cultural historian of twentieth and twenty-first century China, whose research centers around the role of sound in modern Chinese culture. Working at the intersection of cultural history, sound studies, and performance studies, her scholarship examines how media technologies and cultural techniques of sound shaped novel sensory experiences, enacted political and social change, and created new modes of subjectivity in modern China. She received her PhD in Asian languages and cultures from the University of Michigan with a certificate in world performance studies, and currently teaches modern Chinese history, media, and popular culture at NYU Shanghai as a visiting clinical assistant professor.

David Humphrey is an assistant professor of Japanese and global studies at Michigan State University. He is the author of the book *The Time of Laughter: Comedy and the Media Cultures of Japan* (University of Michigan Press, 2023), and his research on Japanese media and digital studies has appeared in journals including *Media, Culture & Society*, the *International Journal of Communication*, and the *Journal of Japanese Studies*.

Eunice Ying Ci Lim is a PhD candidate in comparative literature and Asian studies (dual title) at Pennsylvania State University. Her research interests are in contemporary Southeast Asian and East Asian literature and media, with a focus on non-Mandarin Sinitic languages, language policy, translingualism, and Sinophone-Anglophone intersections. Her book chapter on representations of gendered food consumption in South Korean television dramas (coauthored with Dr. Liew Kai Khiun) is published in the *Routledge Handbook of Food in Asia* (2019). Eunice has also published in the journal *Antipodes* and *ariel: A Review of International English Literature* (forthcoming in 2023). Her dissertation focuses on the intergenerational effects of Singapore's language policies and how contemporary Sinophone and Anglophone literature and media deftly negotiate and respond to these language policies.

Sheng-mei Ma (馬聖美 mash@msu.edu) is professor of English at Michigan State University in Michigan, specializing in Asian Diaspora culture and East-West comparative studies. He is the author of over a dozen books, including *The Tao of S* (2022); *Off-White* (2020); *Sinophone-Anglophone Cultural Duet* (2017); *The Last Isle* (2015); *Alienglish* (2014); *Asian Diaspora and East-West Modernity* (2012); *Diaspora Literature and Visual Culture* (2011); *East-West Montage* (2007); *The Deathly Embrace* (2000); and *Immigrant Subjectivities in Asian American and Asian Diaspora Literatures* (1998). Coeditor of five books and special issues, *Transnational Narratives* (2018) and *Doing English in Asia* (2016) among them, he also published a collection of poetry in Chinese, *Thirty Left and Right* (三十左右).

Mei M. Nan is a PhD candidate in comparative literature at Harvard University. Her research focuses on modern East Asian literature and media. Her articles on Sinophone and Japanese literature and music have appeared in journals and edited volumes, including *Media Mix* (special issue of *Mechademia: Second Arc* 2023), *Transactions of the Asiatic Society of Japan* (2022), and *New Directions in Flânerie: Global Perspectives for the Twenty-First Century* (2021). Her research has been funded by the Social Sciences and Humanities Research Council of Canada, the Japan Foundation, and others.

Lina Qu is an assistant professor of Chinese at Michigan State University. Her research interests focus on modern Chinese women's cultural practices from print to social media and various genres of East Asian food media. She has published on the topics of Eileen Chang, Li Ziqi, social eating livestream, food documentary, and Internet literature. Her work has appeared in *CLCWeb:*

Comparative Literature and Culture, Asiascape: Digital Asia, China Information, and other venues. She is currently working on her first book manuscript, tentatively titled "The Production of Women's Presence: A Genealogical Study of Literary and Cultural Works by Hungry Women in Modern China."

Tze-lan Deborah Sang is professor of Chinese Literature and Media Studies at Michigan State University. Her main research interests are gender and sexuality, documentary, Chinese-language cinemas, and Taiwan studies. Among her major publications are *The Emerging Lesbian: Female Same-Sex Desire in Modern China* (2003) and *Documenting Taiwan on Film: Issues and Methods in New Documentaries* (2012). Her study on the Modern Girl in early-twentieth-century China is forthcoming with Columbia University Press. She is currently finishing a book entitled *Taiwan's Women Documentary Filmmakers: Public Intellectuals and Innovative Artists.* She is also a poet. Her award-winning Chinese poetry collection *Time Capsules* (*Shiguang jiaonang,* 2021) consists of 115 poems composed mostly during pandemic times. Besides Michigan State University, she has taught at Stanford University, University of Oregon, and Hong Kong University of Science and Technology.

Winnie Yanjing Wu is a PhD candidate in the Academy of Film at Hong Kong Baptist University. Her research interests include transnational film and TV, Chinese diaspora, Asian Canadian studies, and migration studies.

Shuwen Yang is a PhD student in the Department of East Asian Languages and Cultures at Stanford University. She is interested in science fiction, online literature, and fandom studies.

Ying Zhu's research encompasses Chinese cinema and media, Sino-Hollywood relations, serial dramas and streaming platforms, and youth digital culture. She is the author of four research monographs, including

Hollywood in China: Behind the Scenes of the World's Largest Movie Market (2022) and *Two Billion Eyes: The Story of China Central Television* (2013), and six coedited books, including *Soft Power with Chinese Characteristics: China's Campaign for Hearts and Minds* (with Kingsley and Stanley Rosen, 2019), *Art, Politics and Commerce in Chinese Cinema* (with Stanley Rosen, 2010), and *TV China* (with Chris Berry, 2019). The recipient of a (US) National Endowment for the Humanities Fellowship, an American Council of Learned Societies Fellowship, and a Fulbright Senior Research Fellowship, Zhu's writings have appeared in prominent academic journals, including *Journal of Cinema and Media Studies* and *Screen* as well as media outlets, including the *Atlantic*, *Foreign Policy*, the *Los Angeles Times*, the *New York Times*, and the *Wall Street Journal*. She reviews manuscripts for major publications and evaluates proposals for research foundations in Australia, Canada, Hong Kong, Sweden, and the United Kingdom. Previously faculty at the City University of New York, Zhu is now faculty at the Hong Kong Baptist University.